CONDENSING AND COOLING

IN

POT-STILL AND PATENT-STILL DISTILLATION

CONDENSING AND COOLING

IN

POT-STILL AND PATENT-STILL DISTILLATION

BY

J. A. NETTLETON

(FELLOW OF THE INSTITUTE OF BREWING)

Author of
The Manufacture of Spirit in the Distilleries of the United Kingdom,"
"The Flavour of Whisky — as influenced by Materials and Processes,"
"Original Gravity," &c., &c.

London
MARCUS WARD & CO., LIMITED
ORIEL HOTJSE, FARRINGDON STREET, E.G.
BELFAST AND NEW YORK
1897

J.A. NETTLETON BIBLIOGRAPHY

1880

"Beer Duty" article, The Times, p.4 col.4, Sept.14.

1881

"A Study of the History and Meaning of the expression "Original Gravity" as applied to Beer, Worts, and Distillers' Wash" (book) The Brewers' Guardian, July 5 through Sept.27, 1881.

1884

Every Brewer His Own Analyst (book).

1887

The Adulteration of Beer by Publicans and Beer Dealers, and its Detection by Brewers, Etc. (book).

1891

"The Preservation of Hops" article.

1892

The Manufacture of Vinegar (book).

"WHISKY" (entry for Chambers' Encyclopaedia, included in subsequent editions).

1893

The Manufacture of Spirit as Conducted at the Various Distilleries of the United Kingdom (book).

1894

"The Flavour of Whisky" article, Transactions of the Institute of Brewing, Issue 7, pp.176-195.

1897

Condensing and Cooling in Pot-Still and Patent-Still Distillation (book).

"Materials, Plant, and Skill" article, The Journal of the Federated Institutes of Brewing, Vol.4, No.4.

1913

The Manufacture of Whisky and Plain Spirit (book).

Other works may exist in trade journals or newspapers as mentioned or similar publications.

"Flavours of Whisky" and "Materials, Plant, and Skill" were originally presented as papers for periodic industry gatherings and feature transcribed post discussion Q&A.

Portions of "Condensing and Cooling.." were first published in the Wine Trade Review, 1896.

2015 PUBLICATION

Aaron Barker Publishing
whiskywheels@gmail.com

ISBN: 9780990907244

2nd Printing

ACKNOWLEDGEMENTS.

The Author's thanks are due to—

(1) T. STEVENSON, Esq., M.D., of the Chemical Laboratory, Guy's Hospital, London, for permission to use extracts from his *Treatise on Alcohol* in the construction of Table XI. of the present work.

(2) Messrs. DUPRE & PAGE, of the Chemical Laboratory, London Hospital Medical College, for permission to utilise their work upon the *Specific Heat of Alcohol* in the construction of Table II. of the present work.

(3) The Proprietor of the *Wine Trade Review*, Eastcheap Buildings, London, for permission to reproduce an important portion of the present work, and which was formerly contributed by him to the columns of that journal.

J. A. NETTLETON,
High Lane, Burslem, Staff.

Feb., 1897.

CONTENTS.

LIST OF TABLES.

INTRODUCTION.

APPRECIATION AND EXPLANATION OF THE PRESENT WORK.

Joseph Alfred Nettleton was a loyal employee of the United Kingdom's Inland Revenue Laboratory, Somerset House, the governmental agency that monitored the brewing, distilling and other industries involving malted grains, from 1875 to around 1883. His name has since been continually mentioned among the legendary authors and books that set the foundation today's whisky (and brewing) industry is built upon such as Henry Stopes, Alfred Barnard, Samuel Morehead and Jameson's "Truths About Whisky". Nettleton's legacy is well-set by the 1913 publication of "The Manufacture of Whisky and Plain Spirit" (recently re-issued in 2009 by Classic Expressions). The year mentioned is important in relation to this book because that grand edition was not the first full publication of that work. The 1913 edition is essentially the second edition of his original 1893 book, "The Manufacture of Spirit in the United Kingdom". That work is an exhaustive dissertation covering the entire whisky and spirit-making process start to finish leaving no scientific stone unturned. But, as he explains in the opening paragraphs of this particular work, that earlier edition was left somewhat incomplete with a noted lack of coverage focusing on Condensing and Cooling and the mechanics thereof. To include such information would have indefinitely delayed the timely publication of "The Manufacture of Spirit". The 1913 book would eventually remedy this shortcoming with the addition of the 40-page "Cooling & Condensing" chapter along with a number of other important updates and additions. The book you hold in your hands is what bridged the gap between the two editions and fulfills Nettleton's promise to conduct a proper and detailed study of distillate condensing and cooling.

When I first encountered the book a few years ago, I had recently become enamored with whiskies that included worm tub condensing in their production process which are generally thought to produce a heavier body and fuller characteristics in a whisky's palate profile (I've since learned there's more to it than just that). I was specifically interested in finding information that directly addressed and supported what I felt were the distinct benefits of worm tub condensation. I had not found much by that time and when I first read the book's title I thought I'd found just what I was seeking. But, I did not. While this book does look at both worm tub and column condensers as applied to either type of still – pot or patent – no judgement is made as to the quality or general characteristics either method imparts. System efficiencies studied across several examples is the only comparison or basis of evaluation made. The book is meant to be an unbiased guide for the spirits distiller to aide in his selection and design of the optimum condensing method given his particular set of operational parameters (still size, availability of water, ambient temperatures, etc.). Suffice to say, it is a dry and technical read but at a different level still fascinating to read how variations in each method can produce the given results.

As mentioned before, the 1913 "The Manufacture of Whisky and Spirit" included a good amount of information found in this book and it would be well understood if you were curious as to what was carried over and what was not. Starting with the bulk of the pages taken from the book, pages 35 through 63 of "Condensing & Cooling" makes up the largest portion of the later book's chapter of the same subject, pages 466 through 488. This section is the ten situational examples Nettleton provides for his demonstrations. Pages 67 to 73 of the present book are added directly after the examples in "The Manufacture of Whisky and Spirit." Plate III from "Condensing & Cooling" is also included with some explanatory detail added while Plates I and II did not make the cut. The section titled "Quantity of Refrigerating Liquor..." found primarily on pages 65 and 66 of this book was also omitted from "The Manufacture of Whisky and Spirit."

Of the sections that were continued into "The Manufacture of Whisky and Spirit", it is clear Nettleton at least performed a cursory review and made changes to improve the clarity of the work. There are examples of his 're-write' found throughout the chapter but to adequately illustrate the nature of the changes, we'll look only at two consecutive paragraphs as read in "The Manufacture of Whisky and Spirit" (p.466, Classic Expressions). Starting at this point, the section titles are different. As found on page 35 of this book:

"EXAMPLES OF CONDENSING AND COOLING ON A PRACTICAL SCALE"

Then, from "The Manufacture of Whisky and Spirit" (p. 466):

Further Technical Remarks upon Condensing and Cooling, and Additional Examples on a Practical Scale.

This appears to be a change to better place this book's material in the context of the new. The opening paragraph in "Condensing & Cooling" reads:

> "The greater portion of the matter contained in the following pages (35 to 76), appeared in the *Wine Trade Review*, during the months from March to September, 1896. The original text has been somewhat amended and extended, and reappears in its new form here, with the kind permission of the proprietor of that journal."

The same paragraph in "The Manufacture of Whisky and Spirit":

> "A portion of these few following pages appeared in the *Wine Trade Review*, March to September, 1896. The original text has been amended and extended, the author having

obtained the kind permission of the proprietor of that journal to re-edit his previous contributions."

As you can see, it's the same content just stated differently. As for the second paragraph of that section in "The Manufacture of Whisky and Spirit," it is entirely new and reads:

> "In patent-still distillation, or, in fact, whenever a liquid, as distinct from vapours, is discharged direct into the worms, certain of the primary and secondary conditions just described do not obtain. And so long as the liquid for distillation, as in a rectifying-column, itself effects the condensation of the rising vapours, so that the final distillates quitting the still proper are only liquids, then boiling-points and latent-heat units are of little concern in the matter of refrigeration."

This new paragraph serves as a link between the new (first six) examples recounted in "The Manufacture of Whisky and Spirit" and the 10 examples that were carried through from this work. Only a slight bit of explanation with no further informational significance. As mentioned before, similar and subtle updates are found through the remaining portions of the chapter. It all seems to be an effort to improve flow and clarity on the reader's behalf.

What you may get from the publication of this book is a greater understanding of the processes and examples Nettleton carried into his later work. Found in "The Journal of the Federated Institutes of Brewing, Vol.3 (1897), this is the best summary of the intentions of the original publication:

> "The subjects treated of in this small work were omitted from the author's volume, *The Manufacture of Spirit*, as being at the time too intricate to be dealt with. The present work may therefore be regarded as supplemental to the previous volume. The contents are as follows:—
>
> Definition and application of the terms specific heat and latent heat. Thermal units. Boiling points. Primary and

Secondary conditions affecting condensation and cooling, and the effects of these in Patent-still Distillation. Practical examples of Distillation and Refrigeration.

The book contains several tables of physical constants used by the distiller.

A list of "Errata" for the author's previous book is appended."

While not an exact replica of the original title (a hardback edition), I attempted to stay as close to the original form as possible while still providing a reader-friendly and affordable edition (my constant goal, book to book). The physical size of the book matches the original as does the format and content from page to page – including the pagination in comparison to the original text. The "Introduction", Nettleton bibliography and any modern copyright or printing information were my only additions; I hope they enhance your experience with this book. Any errors from the original were left in except for one glaring mistake:

The 'heat-units' mathematical equation found on page 65 originally listed the answer as **2,997,300** which is wrong. It is off by a factor of 10x. The correct answer—and the one you'll find at that point in this book—is **29,973,000**. This correction has been carried over to the continuing equations on page 66 and all results and calculations are right.

The electronic edition of this book contains a much simpler formatting and does not replicate the original structure of the printed book but does contain all the same images and information.

Finally, a few biographical notes about Joseph Alfred Nettleton. In addition to re-publishing this book for its first time, I hoped to find a picture or some information about the man himself but what became painfully obvious is that very little record survives from the period of his lifespan – certainly in the case of the early years. However, I did discover a few basics shortly before this publication. Nettleton was born in September, 1851 just outside London in Chelsea. In August, 1875 he began his long association with Inland Revenue as a 'Certified Candidate.' He married his wife, Isabella Redmayne (b.1855), in 1881 at Skipton, North Yorks. The couple had two children, Harold and Edith. Harold would follow somewhat in his father's footsteps and publish many scholarly writings of his own for Universities and Associations in and around London. His work can be found generally credited to 'Harold Redmayne Nettleton' or 'H. Redmayne Nettleton.' JA Nettleton would retire from active publishing after 1913 and spent time in Elgin, Scotland as a consultant to the distilling and brewing industries. Here, he would be visited on or about April 18, 1919 by Masataka Taketsuru, the "father" of the Japanese whisky industry who would also endeavor to create a Japanese translation of Nettleton's 1893 book, "The Manufacture of Spirit" (nonjatta.com/2007/06/unlocking-scotlands-secrets-part-i.html). On October 7, 1939, near Hitchins, Joseph Alfred Nettleton passed away at the well-matured age of 88.

Thanks primarily go to Colleen Card of the Carmel Clay Public Library for her assistance in sourcing the original material. Your local library can be an amazing resource for nearly any area of study and I thoroughly encourage you to engage and support their services often.

Cheers!
Aaron Barker
Carmel, Indiana
April 22, 2015

CONDENSING AND COOLING OF DISTILLATES.

APOLOGY FOR THE PRESENT WORK.

IN the concluding paragraphs of pages 284-285, "The Manufacture of Spirit," there is an admission that a special study of the subjects of Condensing and Cooling, as well as the Construction of worms, worm-tanks, and stills, seemed necessary, in order to convince distillers and distillery engineers of the wide field open for improvements in these respects. The subjects were too intricate to be dealt with at the time that work was being prepared for publication; indeed, the publication would have been postponed indefinitely. Since that date, however (1893), the author has devoted some time to all the questions connected with these difficult but interesting subjects. The following pages are the result of that work.

The present treatise may therefore, in this and other respects, be regarded as supplemental to the previous standard work.

SPECIFIC HEAT.

This term is used in connection with the raising of the temperature of the refrigerating liquor in the worm-tanks, and the corresponding lowering of the temperature of the liquids within the worms; such liquids being low-wines, feints, and spirits. Further, it is used occasionally in reference to the lowering in temperature of the vapours of water, and of the vapours of the different alcoholic liquids just mentioned. The term, too, is equally applicable to solids—to such, for instance, as the iron or wood of the tanks, and the copper of the worms, still-necks, &c. In each case, therefore, it refers to the raising and lowering of the temperatures of these substances during the refrigerating operations.

The following table embraces the Specific Heats assigned by various authorities to all the substances which concern the distiller's operations:—

TABLE 1--SPECIFIC HEATS.

Names of Substances (equal weights of).	Specific Heats.	Authorities.
Water, pure, distilled, temp. 32°F. ...	1°(as standard)	Regnault.
Water, " " average through range, 32°-212°F.	1·005	
Water (ice), average through range, 0°-32°F.	·480	
Water (vapour or steam), temp. 212°F.	·481	Regnault.
Alcohol (ethyl alcohol), Drinkwater's absolute, sp. gr. ·7938 at 60°F. ...	·604	Dupré & Page.
*Alcohol and water mixture, 50 over proof, sp. gr. ·8475 at 60°F. ...	·717	do.
*Alcohol and water mixture, proof strength, sp. gr. ·9198 at 60°F. ...	·910	do.
*Alcohol and water mixture, 50 under proof, sp. gr. ·9671 at 60°F. ...	1·040	do.
*Alcohol and water mixture, 60 under proof, sp. gr. ·9733 at 60°F. ...	1·040	do.
*Alcohol and water mixture, 70 under proof, sp. gr. ·9789 at 60°F. ...	1·030	do.
Alcohol (vapour)	·453	Regnault.
Amyl alcohol (fusel oil constituent), temp. 32°F.	·564	Kopp.
Acetic acid (feints constituent)... ...	·509	do.
Butyric acid " "	·503	do.
Acetic ether " "	·496	do.
Acetic ether (vapour)	·401	Regnault.
Iron	·110	
Copper	·095	

*Calculated from the original experiments of Dupré & Page.

The term *Specific Heat* does not apply directly to the actual temperature or heat-indication of substances, but to the relative quantity or ratio of heat required to raise a given weight of each substance through a fixed range of temperature, and equally to the same relative quantity of heat evolved in lowering the given weight of each through the same range. In this ratio, or relation of quantity, the heat required for raising, or evolved in lowering, a given weight of water is taken as the standard, and quoted as unity, or (1). Water, then, is taken as the standard of comparison for all other liquids, as well as for solids, vapours, and gases. Strictly speaking, and in compliance with the definition found in the text-books on Heat, the specific heat of a substance is the ratio of the quantity of heat required to raise a given weight of the substance through one degree of temperature, with the quantity of heat required to raise the same weight of pure distilled water through the same temperature, the comparison being made when both the substance and the water are at 32° Fahrenheit, or 0° Centigrade.

This ratio, or Specific Heat, strictly holds good only when the substances compared are each raised from 0°C. to 1°C.; that is, from 32°F. to 33°·8F. The amount of heat required to cause an elevation of one degree, as registered by any thermometric scale, varies very slightly with the temperature at which the experiment is made. In the case of water (and of other substances) more heat is required to raise its (or their) temperatures through one degree at the higher than at the lower temperatures. Thus, with water itself, its specific heat taken as 1 at 32°F., rises to 1•005 if averaged for the range of temperatures 32° to 212°F. So with other substances—liquid, solid, vaporous, or gaseous.

The expenditure of heat, its value, or quantity in heat-units, having been fixed as unity for water, that found to be required for equal weights of other substances was ascertained by experiments with the water itself, and in the ways described in the text-books on Heat. The quantity of heat, expressed in heat-units, required by a given weight of water for elevating its temperature one degree, thus becomes a measure or divisor for the heat-units employed in the same work upon other substances.

With almost all substances, liquid, solid, vaporous, or gaseous, the amount of heat required to cause similar elevations in temperature for a given weight of them, or the amount of heat evolved in reducing their temperatures, is less than is required for or evolved by water, consequently in the Specific Heat ratio the unit measure is nearly always higher than the other terns in the ratio; thus—

Specific Heat of Water ... ... = 1.
„ „ Alcohol, •7938 sp. gr. at 60°F. = •604

But there is one remarkable exception to the higher Specific Heat of water which must be referred to now, especially as it affects considerably the subsequent calculations. The Specific Heat of absolute alcohol is much lower than that of water: this has been long known. It was formerly supposed, therefore, that when varying alcoholic mixtures were dealt with, the Specific Heat of such mixtures would be proportional to the relative weights of pure alcohol and pure water present. But in 1869, two chemists, Dupré and Page, proved by careful experiments that, through the whole series of possible strengths, the alcoholic mixtures did not possess the supposed mean specific heats as calculated from the proportional weights of the two constituents. The actual specific heats were invariably found to be higher than the theoretical ones. Further, between the strengths 30 under proof and 95 under proof, it was shown that the mixtures possessed higher specific heats than did water itself. (See Phil. Trans. Roy. Soc., 1869, pages 591, *et seq.)*

The reason assigned for these variations from the calculated Specific Heats is, the probable formation of definite, though unstable chemical compounds of water and alcohol, as distinct from mere mechanical mixtures of the two substances.

The following table embraces some of the results of the investigations of Dupré and Page. The column (E) is chiefly theirs, a few additional strengths having been inserted for the purposes of this work, as have also been the other columns.

TABLE II. – SPECIFIC HEATS OF MIXTURES OF ALCOHOL AND WATER.

Specific Gravities at 60°F. Water=1.	PERCENTAGES OF ABSOLUTE ALCOHOL PRESENT. (Drinkwater's Alcohol, sp. gr. .7938 at 60°F)		Corresponding Proof Strengths (Sikes).	Specific Heats. Water=1.
	By Weight.	By Volume.		
A.	B.	C.	D.	E.
·9912	5	6·20	89°·1 under-proof	1·0150
·9839	10	12·40	78°·3 "	1·0357
·9714	20	24·50	57°·1 "	1·0436
·9576	30	36·15	36°·6 "	1·0260
·9393	40	47·35	17°·0 "	·9680
·9290	45	52·70	7°·7 "	·9419
·9198	49.25	57·05	Proof	·910
·9182	50	57·80	1°·3 over-proof	·9063
·8957	60	67·70	18°·6 "	·8433
·8721	70	76·95	34°·8 "	·7844
·8481	80	85·50	49°·7 "	·7169
·8228	90	93·30	63°·5 "	·6576
·7938	100	100·00	75°·25 "	·6043
·7935			75°·37 "	

As the specific heat of water =1, and that of absolute alcohol (sp. gr. •7938 at 60°F., Drinkwater's alcohol) = •6043, it is evident from the above table that at all intermediate strengths the actual specific heats found by Dupré and Page are higher than the theoretical ones calculated from the respective percentages or proportions by weight of alcohol and water present.

For any intermediate strength not quoted in the foregoing table the appropriate specific heats can be calculated by averaging from the two proximate mixtures above and below respectively. Thus for the intermediate strength 46°•8 under proof, which is half-way between 57°•1 and 36°•6 under proof, $^{(57.1° + 38.6°)}/_2$ = 46°•8 under proof, the specific heat would be the mean of those corresponding to these two strengths respectively, namely, 1•0436 and 1•0260, that is, $^{(1.0436+1.0260)}/_2$ = 1•0348; and so for any other intermediate strength.

It must be borne in mind that substances which have the higher specific heats not only require the expenditure of more heat to raise a given weight of them through one or any number of degrees of temperature, but that in cooling down through the same range of temperature they evolve more heat, and, in fact, they evolve the exact quantities consumed in raising them; hence they require more refrigerating liquor to effect the same apparent work, as indicated by the lowering of temperature.

The specific heats of the liquids water and alcohol, and of the varying mixtures of these, must be known to enable the calculations of the quantities of refrigerating liquor required for cooling down the distillates through any range of temperature to be correctly performed. The specific heats are used in these calculations as co-efficients or co-factors.

It must be remembered that not only liquids have their respective specific heats, but solids and vapours have theirs also. In the latter case, too, the specific heat of water-vapour is higher, though only slightly so, than those of such vapours as alcohol, ether, &c. But as none of the vapours passing over in ordinary free distillation are super-heated, being only at their minimum temperature or normal pressure of vaporisation, no account need be taken of the specific heats of vapours.

Latent Heat.

Latent Heat.—In this work the term latent heat need only be considered so far as it is applicable to evaporation, or more accurately to ebullition, and at the boiling-points, under normal pressure of the two principal liquids, water and alcohol, as well as of the varying mixtures of these. With the latent heat of fusion, that is, of the melting of solid substances, we have nothing to do.

Latent heat is the heat which is rendered latent, is abstracted, or obscured by or in a substance, when such substance, owing generally to its elevation in temperature, and to a series of fixed points for each substance, changes its condition from the solid to the liquid form, or from the liquid to the vaporous form. The attention may be confined for the present purpose to the latter change of state, and though the said change is generally the accompaniment and consequence of elevation of temperature, it is not invariably the case. For any given liquid capable of transformation into the vaporous condition, a constant quantity of heat, varying with the temperature of evaporation or of ebullition, and consequently with the pressure, is required for maintaining a given weight of the vaporising substance in its *new* condition. This amount of heat is not sensible heat, it cannot be registered directly by the thermometer, but during the vaporisation of a liquid a fixed quantity of heat for each substance, and according to the pressure, is always abstracted from the source of heat which causes the vaporisation to occur; or rather the aforesaid fixed quantity of heat is abstracted from the heat already imparted by the original source of heat to the vaporising liquid. Further, when the vapour is again condensed to the liquid form by any cooling process, natural or artificial, the original weight of it, and for any substance at any pressure, renders sensible exactly the original quantity of heat that was formerly rendered latent during the vaporisation. Thus in actual pot-still distillation the alcoholic vapours which enter the worms, on condensing to the liquid form, yield up, or render sensible again, the immense quantities of heat previously extracted from the boiling liquids in the stills, whence they first assumed the vaporous condition. This repayment occurs before the cooling down of the new liquid commences.

Latent heat, as applied to ebullition, is defined in the text-books as the amount of heat, expressed in heat-units, rendered latent during the change of condition of an unit weight of a substance when at its ordinary boiling-point, and at normal pressure, from the liquid to the vaporous state. It must be remembered, however, that ebullition may occur at higher temperatures and higher pressures, and at lower temperatures and lower pressures; also, that evaporation may occur at all pressures and at all temperatures below the normal boiling-point of the liquid, and that in all these cases corresponding losses of sensible heat occur through such heat being rendered latent.

In this sense the latent heat of any weight of a substance is a real quantity or value; that of steam or water-vapour at 212°F. being 536, or, according to other technical authorities, 537°•2 heat-units by the Centigrade scale, which latter are equivalent to 966°•9 heat-units by the Fahrenheit scale, and in each case for 1 lb. of water-vapour under normal pressure. The respective heat-units by the two scales are identical in value, as 5°C. are equivalent to 9°F. Again, the latent heat of a similar weight of alcohol-vapour from alcohol at •7946 sp. gr. at 60°F. is 364 heat-units Fahrenheit.

For the convenience of calculation the respective latent heat-units of the various vapours dealt with in this work can be expressed as a ratio, taking the latent heat units of water-vapour under normal conditions as 1. In this case the latent heat units of the vapour of the above-mentioned alcohol would be in the ratio of 966°•9F. to 364°•3F., and that is as 1 to •377.

The following table shows the respective latent heat units of equal weights of the various vapours, with their corresponding ratios to 1 for water-vapour, at the normal boiling-points of the respective liquids :—

TABLE III. – LATENT HEAT UNITS OF VAPOURS, AND RATIO OF SUCH TO THOSE OF WATER-VAPOUR, AT NORMAL PRESSURE.

Names of Vapours (1 lb. weight of each).	Laten Heat Units, Degrees F.	Corresponding Ratios. (Water-Vapour=1).	Authorities.
Water-vapour (steam)	966°•9	1•	Regnault.
Alcohol vapour	376°•1	•389	Favre and Silbermann.
Do. do. ...	364°•3	•377	Andrews.
Fusel oil(amyl-alcohol)	218°•5	•226	Favre and Silbermann

See also Table on page 261, "Manufacture of Spirit."

HEAT-UNITS.

According to the text-books on Heat, the heat-unit or thermal unit is the amount of heat required to 'raise 1 lb. of water at 0°C. or 32°F. one degree in heat by either scale, namely, to 1°C. or to 33°F. respectively.

In this sense the respective heat-units are not identical in value, as an unit weight of water raised from 0°C. to 1°C. would have to be raised from 32°F. to $33^4/_9$°F. for the same expenditure of heat.

For the present purpose the heat-unit must be measured by the Fahrenheit scale (as it is taken in other technical, as opposed to purely scientific works). The heat-unit, therefore, implies the amount, quantity, or value of the heat necessary for raising 1lb. of water at 32°F. to 33°F.

Strictly speaking, the heat-unit measure cannot be accurately applied at any higher temperature than 32°F., because the quantity of heat necessary for raising the unit weight 1 lb. or any other weight, through one or more degrees of temperature is very slightly greater at all temperatures higher than 32°F. This fact is generally neglected, and purposely so in the text-books, and may well be so here. Otherwise a varying specific heat and heat-unit value would be necessary for every range of temperature. The error introduced by applying the term indiscriminately at any temperature up to 212°F. is so slight, whilst the convenience of using the fixed unit is so great, that in practical or technical work this is always done. As a fact, there are 182 heat-units expended in raising 1 lb. of water from 32° to 212°F., but they are reckoned as 180, corresponding, therefore, to the degrees of temperature between 32° and 212°F. Each such average heat-unit really possesses the mean value of 1.01 standard heat-unit, and this value is also that of the mean specific heat of water through the same range compared with the standard 1 for water raised from 32° to 33°F. But for the reasons above stated, the amount of heat required to raise 1 lb. of water from 32° to 33°F., or from 62° to 63°F., or from 92° to 93°F., will be regarded as identical in value and be called a heat-unit.

In distillation as conducted in the distilleries of the United Kingdom, the temperatures, except in a few special cases, and for private purposes, are always expressed in the Fahrenheit scale, not in the Centigrade or Reaumur. It is, therefore, convenient to do the same thing in the present work. A heat-unit, therefore, always means in this work the amount of heat required to raise 1 lb. of water through 1°F. It means also the amount of heat evolved or abstracted in cooling 1 lb. of water through 1°F.

It might have appeared convenient for working on a large scale to have taken the unit of weight as that of a gallon of water at 62°F., namely, 10 lbs., rather than to have retained it at 1 lb. In that case the number of heat-units to be dealt with in the calculations would have been reduced to one-tenth, and each one would have been reciprocally increased in value. This small advantage would, however, have been gained by the sacrifice of accuracy, inasmuch as 10 lbs. only represent a gallon measure at one temperature, namely, 62°F., whilst the weight of the fixed measure varies from 10·010 lbs. at 40°F. to 9·939 lbs. at 100°F., these being the possible limits within which a refrigerating supply may have to be dealt with.

The following table will enable the heat-units to be calculated correctly, though dealing with gallons by measure within the range of temperatures.

TABLE IV – WEIGHT PER GALLON MEASURE OF WATER AT VARYING TEMPERATURES.

Temperatures.	Corresponding weights per Gallon measure	Temperatures.	Corresponding weights per Gallon measure
40° F.	10·010 lbs.	62° F.	10·000 lbs.
45° ,,	10·008 ,,	70° ,,	9·991 ,,
50° ,,	10·007 ,,	80° ,,	9·979 ,,
55° ,,	10·004 ,,	90° ,,	9·966 ,,
60° ,,	10·002 ,,	100° ,,	9·939 ,,

BOILING-POINTS.

The temperatures at which any of the alcoholic liquids (wash, low-wines, feints, and spirits) boils have to be considered in estimating the amount of cooling liquor required. At any fixed pressure, and under any other conditions whatsoever, the temperature of the vapour given off by a similar boiling liquid is identical and constant, and is generally identical with that temperature at which the liquid is boiling. If, by the adoption of a specially-shaped vessel, or of one made of a particular material, or if salts or other solids be dissolved in the liquid, its boiling-point may be raised or lowered, but the vapours driven off will be at a constant temperature if only a constant pressure be maintained. This remark applies to homogeneous liquids, not to mixtures of liquids the constituents of which have varying boiling-points. Thus with wash boiling at 200°F. at any given time, the vapours simultaneously evolved will be at 200°F. also. But in the case of all distillery liquids undergoing distillation the boiling-points gradually rise towards 212°F., and in the case of wash perhaps a little above that temperature. The most volatile constituent begins to quit the liquid at the initial boiling-point, and passes into the worm at that particular temperature. Any portion of it that is condensed on its way to the worm has to be re-vaporised. The residue in the still, as distillation proceeds, becomes gradually richer in water, in acid, in oil, and, if the liquid be wash, in unfermented wort extract. For this reason the temperature of a charge of wash will gradually rise with the boiling-point from 198°F. to nearly 212°F. before all the alcohol is driven off. The vapours successively entering the worm during the progress of the distillation will thus vary in temperature from 198° to 212°F., of which limits 205°F. is the mean temperature. What yet remains to be known is the respective weights of the vapours which entered the worm at any intermediate temperature, for the nearest mean temperature of vaporisation of a given mixed quantity of vapours can only be ascertained by averaging from the products of a series of such weights at a series of temperatures. But it is impracticable to ascertain these details in actual work, and all that can be done is to take the mean temperature from the first and final boiling temperatures, irrespective of weights or volumes of vapours differently constituted, and only collected when condensed again. It must be for the moment assumed that the boiling operation from beginning to end proceeds at the normal atmospheric pressure, though it probably takes place at a lower pressure throughout, and therefore at a lower temperature, as explained on page 43. This holds especially with furnace-heated stills, or steam-jacketed or steam-coiled stills, but not to the same extent when naked steam discharges into the boiling liquid, for then the super-heated and super-pressed steam tends to equalise the internal pressure. It would be easy to ascertain the pressure at which the vapours are evolved in any case by means of a pressure-gauge in the neck or bend of the still.

In patent-still distillation a corresponding elevation of the temperature of the wash takes place during boiling, for the residues falling down to the lowest chamber of the analysing column become hotter as they approach and mingle with the raw steam at its full pressure.

The following table embraces details relating to the boiling-points of the principal liquids whose vapours pass over into the worms in pot-still distillation. In patent-still distillation the condensed vapours or liquids only pass into the worms, except a little ether vapour from the top of the rectifying column.

TABLE V. – BOILING POINTS.

Names of Liquids or Mixtures of Liquids.	Temperatures at Normal Pressure of Air (760 mm.) (90 In.)	Authorities.
Water	212°F.	
Wash, containing 1 per cent. Proof Spirit or ·45 per cent. Alcohol by weight ...	211°F.	Gröning.
Wash, containing 2 per cent. Proof Spirit or ·95 per cent. Alcohol by weight ...	209°F.	do.
Wash, containing 5 per cent. Proof Spirit or 2·20 per cent. Alcohol by weight ...	209°F.	do.
Wash, containing 10 per cent. Proof Spirit or 4·55 per cent. Alcohol by weight ...	203°F.	Ure.
Alcohol, absolute (Drinkwater's), ·7938 sp. gr. at 60°F., 75·25 over-proof, 100 per cent. Alcohol by weight	172°·2F.	Dupré & Page.
Alcohol and Water Mixture, 63·5 over-proof, 90 per cent. Alcohol by weight ...	173°·3F.	do.
Alcohol and Water Mixture, 49·7 over-proof, 80 per cent. Alcohol by weight ...	174°·8F.	do.
Alcohol and Water Mixture, 34·8 over-proof, 70 per cent. Alcohol by weight ...	176°·0F.	do.
Alcohol and Water Mixture, 18·6 over-proof, 60 per cent. Alcohol by weight ...	177°·6F.	do.
Alcohol and Water Mixture, 1·3 over-proof, 50 per cent. Alcohol by weight ...	179°F.	do.
Alcohol and Water Mixture, proof, 49·24 per cent. Alcohol by weight	179°·2F.	(Interpolated.)
Alcohol and Water Mixture, 7·7 under-proof, 45 per cent. Alcohol by weight ...	180°·5F.	Dupré & Page.
Alcohol and Water Mixture, 17·0 under-proof, 40 per cent. Alcohol by weight ...	181°·5F.	do.
Alcohol and Water Mixture, 36·6 under-proof, 30 per cent. Alcohol by weight ...	184°F.	do.
Alcohol and Water Mixture, 57·1 under-proof, 20 per cent. Alcohol by weight ...	188°·6F.	do.
Alcohol and Water Mixture, 78·3 under-proof, 10 per cent. Alcohol by weight ...	196°·7F.	do.
Alcohol and Water Mixture, 89·1 under-proof, 5 per cent. Alcohol by weight ...	202°F.	(Interpolated.)

For Boiling Points of othe Wash and Feints constituents, see pages 358 to 360, "Manufacture of Spirit."

The actual boiling-points of alcohol and water mixtures are always several degrees of temperature (4 to 16, according to strength) lower than the theoretical boiling-points calculated from the average of those of the two constituents, along with the proportions by weight of these present in the mixtures. Thus water 212°F. b.p., alcohol 172°•2 b.p., average b.p. for 50 per cent. weight of each in mixture 192°•1F, actual b.p. only 179°F.*

Primary Conditions which affect the Condensing and Cooling of Distillates, and which govern the Quantities of Refrigerating Liquor required therefor.

The following conditions must always be considered in calculating the quantities of liquor required for the worm-tanks: —

1st. The temperatures of the boiling-points of the liquids in the several stills, or, what are practically the same, the temperatures of the vapours entering the several worms. The mean temperatures obtaining during distillation are taken.

2nd. The latent heats of the mixed vapours entering the several worms, the mean values of the latent heat-units (or of their ratios to water-vapour) of the mixed vapours being taken.

3rd. The specific heats of the liquids condensed from the vapours and passing through the worms, the meat) specific heat of each separate distillate being taken.

4th. The temperature of the supply of refrigerating liquor.

5th. The temperatures of collection of the several distillates in the receivers.

*(For definitions and explanations of other useful physical terms, and for other tables of physical phenomena, see Chapter VII., "Condensing and Cooling of Distillates," in the author's standard work, *The* Manufacture of *Spirit in the Distilleries of the United Kingdom.)*

6th. The temperatures at which the refrigerating liquor in the several worm-tanks, or compartments thereof, is allowed to overflow after having done its work. This condition includes the important question of fractional condensing and cooling.

Secondary Conditions affecting the Condensing and Cooling of Distillates, and which, govern (in a less degree) the Quantities of Refrigerating Liquor required therefor.

There are many minor conditions which affect both the actual and the calculated quantities of refrigerating liquor, but which cannot be allowed for in a general calculation. The necessary corrections they involve can be introduced in any special case. With respect to them it may be said :—

(a) They differ in many ways at every distillery, and their effects cannot be generalised or averaged approximately.

(b) They also vary slightly with each distilling operation at any one distillery.

(c) Several of them are of an intricate character, and, if considered, would render the calculations very difficult.

(d) The corrections they would effect in the calculated quantities of refrigerating liquor must in nearly all eases reduce the gross estimate. On a small scale of working they would save more proportionately than on a larger one. The extent to which they affect actual and estimated quantities may be judged of from the few following pages.

The following list embraces these secondary and varying conditions:-

1st. The variations in the alcoholic strengths of the wash, low-wines, and feints tendered for distillation in succeeding periods. These variations involve variations in the temperatures of the boiling-points of such liquids, consequently in the temperatures of the vapours. The latent heat-units of the vapours must therefore vary, so must the specific heat of the distillates when these are condensed. But the diversities introduced in this way would be small if a regular system of brewing and distilling from period to period be followed out. When any permanent change is introduced, suitable corrections can be made by referring the new conditions to the Tables II., III., and V. in this work.

2nd. The exposure of the surfaces and sides of the refrigerating tanks, and the consequent evaporation and cooling. The extent of the cooling so effected must vary continually for the same plant, and varies considerably for different localities. An exposed position to the sun's rays during a hot season, or in a hot climate, may cause a waste of liquor through heating; but in the usual way in our northern latitudes, exposure means an extra cooling effect, and therefore economy in liquor. The extent of this cooling effect depends upon the temperature of the overflow water, that of the surrounding air, the exposure of the liquor and the tank to draughts of air, to wind, rain, and sun.

That evaporation alone carries off a large quantity of liquid, and of heat previously absorbed by it from the hot worms, is evident from the following table, which has been calculated for application here from Dalton's original experiments. It is not the quantity of water carried off that is of any importance, but the cooling effect produced by the vaporisation of such water through the transference of sensible to latent heat.

TABLE VI – QUANTITIES OF WATER AT VARYING TEMPERATURES, CARRIED OFF BY EVAPORATION, PER HOUR, PER SQUARE FOOT OF SURFACE.

Temperatures.	In Calm.	In Breese.	In High Wind.
55°F.	580•26 grains.	742•12 grains.	910•09 grains.
65°F.	800•14 ..	1029•19 ,,	1258•81 ,,
75°F.	1104•71 ,,	1429•27 ,,	1746•88 ,,
85°F.	1502•56 ,,	1982•04 ,,	2455•41 ,,

Let the above data be applied to an actual operation, in which it is supposed that the overflow surface of the liquor in the tank is exposed to a light breeze, that the temperature of the overflow is 85°F., and that the exposure lasts for a period of 48 hours, during which time distillation is proceeding continuously, and hot vapours or liquids are passing through one or more worms in the tank. If the surface dimensions of such a tank be 36 feet by 12 feet, there would be 432 square feet of exposed surface of overflowing liquor. Calculating the quantity of water carried off by evaporation under these circumstances, and in standard gallons, the total works out as follows : —

$(1982.04\text{grains} \times 48 \times 432)/_{70,000}$ = 587 gallons. This means $12^{1}/_{5}$ gallons per hour's exposure, and the conversion of a large quantity of sensible heat into latent heat. The heat-units extracted for retaining the vapours in their new form must come from the sensible heat of the overflowing water, and consequently the residual water near the surface, as well as that slowly overflowing, must be colder by the same number of heat-units. This action is equivalent to the requirement of less refrigerating water. And if the latent heat of aqueous vapour evolved at any such temperature as that possessed by the overflow liquor were as high or as low as that for steam produced at 212°F., the latent heat-units extracted would, in accordance with the foregoing example, amount to 965°F. x 587 x 10=5,664,550.

The evaporation does, under any circumstances, produce a cooling effect on the liquid and on the surrounding air. The laws governing the rate and quantitative value of evaporation in this case are generally included in the four following conditions :—

The area of the surface of exposure.
The temperature of the exposed liquid.
The state of dryness or moisture of the surrounding air

The rate at which the superjacent air is renewed and removed. In damp, cold weather, when the air is already quite or nearly saturated with water, it cannot render any assistance in removing any more from a wet surface, nevertheless cooling may be effected by the slower power of radiation.

The great service rendered by revolving fans or other air-propellers placed over open coolers at breweries and distilleries, in dissipating the vapours as soon as these are formed upon the surface of the hot wort, could almost equally well be utilised above the surfaces of refrigerating tanks, and especially when the overflows are slow, but at the higher temperatures usual in fractional refrigeration. One benefit resulting from the use of wide but comparatively shallow tanks in such a system is evident. It is undesirable to board the worm-tanks over, unless they are in a place hotter than that of the overflow liquor; nor to erect them in covered places, unless these are cooler than external positions.

3rd. Radiation of heat from the warmed refrigerating liquor surface, and from the tank side to the surrounding air, so long as the latter is colder than the former. In hot climates it is possible that heat might be absorbed this way from the surrounding air instead of radiated towards it, especially in an exposed position under a mid-day tropical sun. The amount of heat so radiated, or absorbed, as the case may be, depends chiefly upon the difference in the temperature between the tank and its contents on the one side, and the surrounding air on the other. There is *a* constant struggle going on for equalising these temperatures. The value in heat-units of the heat lost or gained this way cannot readily be estimated. Shallow tanks with large surfaces are affected most, and the loss of heat is large when a high temperature of overflow occurs in *a* tank placed in *a* cold position, or during frosty weather.

4th. The shape, position, and construction of the still, still-furnace, head, neck, and worm, affect the temperature of the vapours entering the worm, consequently they affect the quantity of refrigerating liquor wanted. The depth of liquor in the still, the distribution of the furnace heat or of the steam heat, the point of concentration to which distillation is continued, the use of attemperating coils or jackets within the still-neck, the exposure of the still-neck and first bend of the worm to the open air, the connections of the copper lengths in neck and worm, whether by soldering or by non-conducting joints—all these conditions affect either the amount of heat conducted during distillation by the copper towards the worm-connection and along the worm itself, or the heat of the vapours themselves.

5th. The weight of the worm, independent of its distribution in one or more tanks. An example of the deduction to be made in the estimate of the quantity of the refrigerating liquor, owing to absorption of heat by worm and tank, is given on page 276, *Manufacture of Spirit*. In that example the saving amounted to 102 gallons raised through 20°F., or to 204 gallons raised 10°F. The following is another representative case in fuller detail :—

A wash-still worm of copper throughout. Total length, 450 feet, average diameter 7 inches, therefore of average circumference 22 inches (1 ft. 10 in.); average thickness, $^{1}/_{16}$ inch. The worm thus contained 825 square feet of surface, and as one square foot at $^{1}/_{16}$ inch thickness weighs 2•9 lbs. approximately, the total weight of the worm is 2,392 lbs. The specific heat of copper is •095 (water=1), so that 2,392 lbs. of copper are equivalent in heat capacity to 227 lbs. of water. The worm, as in the example given on page 43, rises during refrigeration from an original temperature of 45°F. (that of the supply liquor) to 80°F., the mean temperature of the water in the tank which overflows at 115°F. Hence the whole weight of the copper composing the worm is virtually raised from 45' to 80°F., or through 35°F. of temperature. The heat-units consequently retained by the mass of copper amount to 227 x 35°F.=7,945, so that this quantity of heat in heat-units may be deducted from the gross total of heat-units which has to be disposed of in condensing and cooling the low-wines coming from the wash-still. A gallon of liquor at 45°F. weighs 10•008 lbs., so that 227 lbs. are equal to 22•7 gallons, and they are raised 35°F. each, or 1589 gallons raised through 5°F. This may seem to be but a small correction upon the gross quantity employed, but the deduction could be more than 'doubled if a similar correction were applied for the other two smaller worms of the low-wines still and feints-still respectively. Probably the total correction would not exceed 50 gallons raised through 35°F. on each complete period estimate, and would be of small account in a large operation. The correction would be further augmented, however, if the absorption of heat by the iron or wood of the containing tanks were considered. The iron may be taken as possessing a specific heat of •113.

6th. The rate of distillation necessarily involves a corresponding rate of supply and overflow to and from the tanks. A more rapid supply usually necessitates a higher water-pressure from the mains. There is then more risk of the supply-water rising up in currents instead of distributing itself around and between the worm-coils, and of its escaping at the overflows without having abstracted the maximum amount of heat in its course. The increase of the rate of distillation means also the concentration of the heat as to quantity in a given time, and this condition, in several ways connected with conduction and radiation of heat, evaporation during exposure, and convection in the refrigerating liquor, involves the expenditure of additional quantities of liquor.

7th. A number of details as to various corrections and circumstances connected therewith are embraced in the catalogue (1) to (7) on page 53 of this work. The most important of these details, and one which causes a great waste of precious liquor, is that of the external and internal cleanliness of the worms. The accumulated deposits form a powerful non-conducting substance, thereby preserving the heat and the temperature, and nullifying the effect of cold water.

8th. In support of the conditions described above in the second heading, page 27, the following table, showing the tensions of aqueous vapours at a series of temperatures, will serve to give some idea of the amount of evaporation constantly taking place in dry air, or unsaturated air, at temperatures below the ordinary boiling-point. As previously explained, the importance of this phenomenon does not rest in the quantity of water evaporated, but in the corresponding abstraction of sensible heat and its conversion into latent heat. With fractional refrigeration, and overflows at high temperatures, the number of heat-units so disposed of and the reciprocal cooling effect must be very great during many hours of exposure.

TABLE VII – TENSIONS OF AQUEOUS VAPOUR AT TEMPERATURES BELOW THE NORMAL BOILING POINT. REGNAULT.

Temperatures.	Corresponding Tensions or Vapour Pressures in millimetres of Mercury.	Temperatures.	Corresponding Tensions or Vapour Pressures in millimetres of Mercury.
50°F.	9•165	113°F.	71•391
59° „	12•699	122. „	91•982
68° „	17•391	140°„	148•791
77° „	23•550	158° „	233•093
86° „	31•548	176° „	354•643
95° „	41•827	194. „	525•450
104° „	54•906	212°„ (B.P.)	760•100

9th. Heat of Combination. When alcohol and water are mixed together, the mixture rises in temperature immediately. The greatest rise in temperature is noticed when about 30 parts by weight of absolute alcohol are mixed with 70 parts by weight of water. In such a case the corrected rise of temperature is 48°F•4, though the observed rise is much less, for all the time—a minute or two—that the mercury is rising, cooling is also going on. Hence the heat-units evolved during the mixing transaction are higher in number than the mere observation of the rise in temperature would indicate, and in any case the heat-units must eventually be absorbed by some cooling agency.

If, during distillation on the usual scale, the vapours of alcohol and water were once really separated, and therefore had to be reunited upon condensation, the heat-units of combination would be enormous. They might then be regarded indeed as a vast reserve of potential heat similar to that of latent heat, to be rendered sensible when condensation occurred. But it is improbable that the respective vapours are isolated. Nevertheless, as the distillates when again condensed are at different strengths throughout the process, admixture of stronger and weaker fractions must be continually occurring in the worms and in the receivers, although the admixture in any case is only a partial one, for the distillates always form in different layers or strata of varying strength in the ordinary receivers, and much agitation is required to ensure a proper mixture of large bulks. It is impossible, except by direct observation and experiment, to assign any value in heat-units to combinations taking place under every variety of circumstances connected with bulks, strengths, and more or less partial mixing, &c.

10th. The purity of the refrigerating liquor, and the filtration of the low-wines and feints prior to re-distillation. These two conditions, though quite unconnected, both control to an important extent the necessary quantities of refrigerating liquor.

As to the water. Where this is obtained from surface supplies, such as natural ponds, small lakes, stagnant streams, the water frequently, for a portion of the year at least, abounds in animal and vegetable life, and to such a large extent that the worms become thickly coated with an impervious deposit. So, too, when the stream or other supply is polluted with mineral matter to an unusual degree. These deposits collect upon the worms, and form about the best non-conductor of heat that could be contrived. But this is exactly what is not required, for the conduction of heat being rendered so difficult makes it obligatory to employ far larger quantities of cold water than would otherwise be required. It is desirable, therefore, either to remove the deposited matter periodically, or to take such precautions as will remove at least all the grosser and heavier impurities from the water. The alternatives of rough filtering or of subsidence suggest themselves as remedies, and they can both be effectively carried out on the large scale, and in various ways which need not be dwelt upon here.

As to the filtering of the low-wines and feints from their oily and mineral impurities. Such methods are frequently in vogue, and give much satisfaction, for reasons connected with flavours and purity of the final spirit. But in this place the value of filtering the distillates, and the consequent removal of the oil, rests in the fact that the worms are kept cleaner inside, that conduction of heat is less interfered with, and that refrigerating liquor is saved.

In patent-still distillation by the Coffey still, and in any modification thereof, and in fact whenever a liquid as distinct from vapours is discharged direct into the worms, the conditions, Nos. 1 and 2, Primary Conditions, page 25, affecting the quantities of refrigerating liquor do not obtain. And so long as the liquid for distillation on its way to an analysing column or chamber, but passing first through a rectifying column, effects complete condensation of all vapours, so that the distillates delivered are only liquids, then boiling points and latent heat-units are of no concern in this matter of refrigeration. The other conditions, primary and secondary, do obtain, except some of those scheduled in beading 4, page 30.

Examples of Condensing and Cooling on a Practical Scale.

The greater portion of the matter contained in the following pages (35 to 76), appeared in the *Wine Trade Review*, during the months from March to September, 1896. The original text has been somewhat amended and extended, and reappears in its new form here, with the kind permission of the proprietor of that journal.

It must have often occurred to distillers and distillery engineers, that various improvements which could effect considerable economy in time and in fuel, in the quantity of refrigerating liquor, and in the gigantic size of the ordinary worms, must be well worth attempting, not only when the Coffey still is employed, but more especially at pot-still distilleries, at which high qualities and safe maturing conditions are eagerly looked for. Although working under the present imperfect conditions, many distillers succeed, fairly consistently, in producing a quality of whisky which leaves nothing to be desired in that particular respect, yet, at the same time, the great trouble entailed in so doing cannot be overlooked, involving, as it does, the repeated distillation of several fractions of the whole bulk of the crude distillates, especially if the final product be intended to take its stand as a self-whisky, and not meant merely for blending. The wonder is, indeed, as regards pot-still spirits generally, that the quality reaches so high a standard. This consummation, as is well-known, is brought about mainly by two facts—namely, the use of high-class materials and the skill of the operatives. But these satisfactory results are only attained, as above indicated, at what must be regarded as an extravagant outlay of labour, time, fuel, condensing liquor, and plant.

There are cases, no doubt, where two, if not more, of these items may be considered as of small account—as, for instance, in localities where fuel is extremely cheap, water abundant during the season, economy in labour a point not thought of, and time of no account whatever. Such favourable conditions are enjoyed, more or less, in special districts, but they are fast disappearing before the flowing tide of competition.

There are many practicable ways in which economical improvements can be effected during distillation, and these are so interwoven that it is difficult to select one for demonstration without introducing several others. Let us instance the quantity of condensing and cooling liquor. Sufficient must be forthcoming at a Lowland or at an Irish distillery for condensing the vapours and cooling the distillates from the successive charges of three distinct distillations in as many stills. In the Highland distilleries the quantity of liquor required must be considerably less, probably by 15 per cent., owing to the collection of more spirits direct from the strong low-wines. Now comes a practical question—namely, is it preferable to save some space and some initial expense by erecting only one worm-tank for the three worms, or to erect separate worm-tanks of appropriate sizes for each still. This question has never been discussed on its merits, the decision in any new case being usually controlled by the general practice of the distiller or of his neighbours, whilst the engineer prefers to adhere to one method of worm-tank construction for the varying necessities of pot or patent-still distillation, without considering the question of water supply. Confining ourselves just now to pot-still distillation, we ought to recall to our minds precisely what it is that is required to be done. Now, looking at the enormous bulk of vapours coming from the wash-still as well as from the low-wines-still, and remembering that a third or final distillation of the collected liquids has to follow on immediately, it is obviously unnecessary to cool the first two distillates below a temperature which will prevent the loss of spirit vapours during collection and removal for redistillation; whilst in regard to the final distillate, a lower temperature of collection is necessary, for a far stronger alcoholic liquid is being then dealt with. It is presumed that an efficient testing safe is employed, the sampling pipes in each case being conducted through a miniature condenser. By restricting the cooling of the two former and larger distillates, not only is the quantity of refrigerating liquor greatly reduced, but the time and the fuel required for immediately boiling up again are proportionately restricted. It will be shown further on what a saving of liquor may thus be effected.

There are in vogue three variations of the condensing operations. The first one, sanctioned by long custom and found in application at many old distilleries, involves the erection of three worm-tanks; the second—a more modern idea—requires only one tank, all the worms passing through it; the third variation comprises several modifications. Thus there may be one large tank, divided into three independent compartments of appropriate size and with separate supplies and overflows; or the worm of any of the three stills (generally that of the whisky still), after leaving the principal tank, may pass through a supplementary tank, so as to complete the cooling to the temperature required.

Many other important variations in the construction of worms and worm-tanks must be discussed, in reference to the quantity of refrigerating liquor required, after this main point is disposed of.

Take first a representative case where the weekly production of wash for distillation is 50,000 gallons, say of original gravity 45, attenuated to water. Such wash would contain almost exactly 45÷4•5=10 per cent. of proof spirit, allowing 4½ degrees of attenuation for the production of 1 per cent. of proof spirit. The first distillation produces low-wines, which all pass through one worm, and if they were all collected in one receiver at 60 u.p., or 40 per cent. proof spirit (average strength), would involve in this or any case the collection of 12,500 gallons of this first distillate. The second distillation* involves the collection of 4,762 gallons of weak and strong feints passing through one worm, and if all collected in one receiver would possess an average strength of 5 o.p., or thereabouts. The third distillation yields spirits and feints, of the latter weak and strong fractions, the spirits collected at, say, 44•5 o.p., and the feints at an average strength for the weak and strong fractions at proof. The respective bulks of spirits and of strong and weak feints together are practically as 3 to 4, say 1,800 gallons of spirits and 2,400 gallons of strong and weak feints, the total bulk being for this third distillate 4,200 gallons.

The work required to be performed with the help of the three worms is, therefore, as follows (redistillation of depending feints omitted) : —

1. The wash-still worm must condense the vapours and cool the liquid distilled from 50,000 gallons of wash, say 12,500 gallons of distillate.
2. The low-wines-still worm must condense the vapours

*If the weak feints and the weak low-wines, separately collected from the strong runnings of each, be returned (as is very likely) to the low-wines still, whilst the strong runnings of each pass to the spirit-still, the following calculations will not be affected as comparisons of the total quantity of refrigerating liquor required in the several methods of condensing and cooling, for in each example what one worm escapes is merely taken up by the other. In calculating absolute quantities for each worm, these quantities and the totals for three worms in a series may be restrict.' or increased by varying the method of collection.

from 12,500 gallons of low-wines and cool the 4,762 gallons of liquid so formed.

3. The spirit-still-worm must condense the vapours from 4,762 gallons of strong and weak feints and cool 4,200 gallons, the total of three fractions of liquid therefrom.

As will be shown forthwith, the necessary quantity of refrigerating liquor required in any method of collecting the distillates must vary considerably, according to the higher or lower temperature of the supply, though far more, in fact chiefly so, according to the mean temperature at which the overflow liquor is permitted to escape from the worm-tanks. In endeavouring to take a representative case for the first example let it be supposed that the season's supply of refrigerating liquor can be obtained at a mean temperature of 45°F., and that such a moderate temperature as 62°F. be deemed low enough for the spirits collecting in the receiver. But as all or the chief portion of the low-wines when collected is intended for immediate re-distillation, as well as that portion of the feints not intended .to be kept over until the next period, these liquids certainly need not be cooled below 80'F., except those portions of either intended for keeping over.

It is now necessary to agree upon several important factors as in certain examples detailed on pages 274-284, *Manufacture of Spirit*. Besides, there is the highly valuable work of Dupré and Page to help us here (Philosophical Transactions Royal Society, 1869, pages 591, *et seq.)*.

1. As to the low-wines collected at 60° under proof, and, therefore, of specific gravity •9733 at 60°F., and hence practically weighing 9•733 lbs at the same temperature. The specific heat of such a liquid is 1•0456 (water = 1). And as the mean

temperature of the low-wines vapours on first entering the-worm of the wash-still is practically identical with that of the mean boiling-point of the wash of the alcoholic strength previously quoted (90 under proof), such vapours are first introduced at 208°F. approximately. It must be remembered here that pot-still distillation is always carried on under a partial vacuum in consequence of the condensation taking place in the worms, so that the temperature of the boiling liquids is lowered in all cases, and must be below that of an ordinary open boiling as applied to the respective alcoholic liquids dealt with. The next factor is the mean latent heat of the vapours to be condensed, and which, in this case, may be taken as '9 (water vapour = 1), or, if expressed in thermal units for a given mass, would be 965°F. x •9 = 868°F•5 for low-wines vapour 60 under proof (18•5 per cent. weight of alcohol).

2. As to the feints collected from the second distillation and at a mean strength of 5 o.p. for the weak and strong fractions. Such a mixture would have a specific gravity of •9137 at 60°F, and would weigh approximately 9•137 lbs. per gallon at the same temperature. Its specific heat would be slightly less than that of water, and would be, in fact, very nearly •9 (water vapour =1). Next, the mean boiling-point of the vapours from the low-wines still now entering the worm, and being condensed into feints, would be the mean between the boiling-point of an alcoholic mixture 60 under proof and that of water—that is, the mean of 190° and 212°F.—namely, 201°F.—without allowing for the further reduction in the boiling-point caused by the partial vacuum. Further, the mean latent heat of the mixed alcoholic vapours collected, when condensed in the various feints receivers at the average strength 5 o.p., would be that for the vapours of an alcoholic liquid containing 52 per cent. by weight of pure alcohol and 48 per cent. of water, and as the latent heats of pure alcoholic and of aqueous vapours are respectively as •38

and 1, their mean latent heat would be •69, say •7, or expressed in thermal units per pound mass, 965°F. x•7=675°•5F.

3. As to the spirits and feints produced in the third distillation, collected in the spirit receiver and various feints receivers, at an average strength of 19•3 o.p. Such a mixture would have an average specific gravity at 60°F., of •8948. Consequently a gallon of it, at the same temperature, would weigh very little over 8•948 lbs. The mean specific heat of the mixture would be •84 (water = 1). The mean boiling-points of the liquids collected at this strength from the original mixed feints of the second distillation (at 5 o.p.) would, without the extra reduction due to the partial vacuum, be the average between the boiling-point of an alcoholic liquid at 5 o.p. and of water, or the mean of 177° and 212°F.—namely, 195°F.—so that the mixed vapours entering the spirit-worm would be at this average temperature. Finally, the mean latent heat of the vapours entering this worm, and which, when condensed and collected in the spirit receiver and various feints receivers at 19•3 o.p. average strength, would be proportionate to the respective percentages of pure alcohol and of water present in such vapours. Spirit at 19•3 o.p. contains 60•35 per cent. by weight of alcohol and 39•65 per cent. of water, and, applying this ratio to the respective latent heats of vapours of pure alcohol and of water—namely, •38 and 1—the average works out at •63. Expressed in thermal units per given mass the number would be 965°F. x •63 = 607°•9F.

For convenience, the principal data for use in calculating the required quantities of water in the following nine examples are collected in the annexed tabular statement.

TABLE VIII.

Quantities collected in Receivers	LIQUIDS.	Temperatures.	Specific Gravities at 60°F.	Weights per Gallon (approx..).	Specific Heats.	Boiling-Points (normal pressure) of Liquids producing these Distillates.	Actual Latent Heats of Vapours condensed (Heat-units per 1 lb.).	Ratios of Latent Heats of these Vapours.
(Gals.)				lbs				
	Refrigerating Liquor Supply at ..	45°F	1·0008 at 45°F.	10·008	1.	212°F.	965°F.	1
12,500	Low-wines collected at ..	45°F.	·9733	9·733	1·0456	208°F.	868°·5F.	·9
4,762	Feints from the Low-wines collected at ..	45°F.	·9137	9·137	·9	201°F.	675°·5F.	·7
4,200	Spirits and Feints from the Spirit-still, collected at..	45°F..	·8948	8·948	·84	195°F.	697°·9F.	·63

For each liquid, 1 lb. of it may be taken as the unit weight for calculating the thermal units, but as the specific gravity at 60°F. of each liquid is known, the weight of one gallon of each at 60°F. may more conveniently be used in these calculations. Thus, the cold-liquor supply at 45°F. weighs 10•0086 lbs. per gallon, and this weight must be used in the divisor.

Examples, on a working scale, illustrating methods of condensing and cooling the distillates, and the means of calculating the required quantities of water therefor, can now be furnished.

1st Example—Pot-still distillation. Condensing and cooling in three independent refrigerating tanks or worm-tubs.

This example represents the oldest and best-known practice.

(a) Low-wines from the wash-still. The condensation and cooling of two fractions, strong and weak, of low-wines to the temperature 80°F., require a quantity of water which may be calculated as follows (small corrections for the loss of heat due to the size of the worm, and worm-tank, to evaporation from the exposed surface of the liquor in the tank, and to radiation being omitted) :—

The 12,500 gallons of weak and strong low-wines, previously alluded to, weigh on the average 9•733 lbs. per gallon, or slightly over, at 60°F. The weight of the low-wines is thus 12,500 x 9•7331bs.=121,662•51bs. The heat-units to be absorbed in the condensation alone from the vaporous to the liquid form are, in this and in all cases, several times as heavy, usually five or six times, as those to be absorbed in the subsequent cooling. In this *case* they amount to 12,500 x 9•733 x 965°F. x •9= 121,662·5 x 868°•5F..105,663,881•25 units.

To these must be added the heat-units of the cooling down to 80°F.—namely, 12,500 x 9•733 x 1•045 x (208 — 80) = 121,662•5 x 1045 x 128=16,273,576 units.

Hence the total number of heat-units to be absorbed in the combined condensation and cooling amounts to 121,937,457.

The cooling and condensing are effected, as has been stated, with liquor supplied at 45°F. The mean temperature of such liquor whilst in the tank or tub must be 80°F. exactly, so that, as the liquor flows in at 45° F., it must flow out at a little above an equal rate at 115°F. Each gallon of the supply-liquor weighs 10•008 lbs., and carries off as it flows away 10•008 x 70=700•53 thermal units derived from the distillate in the worm. The quantity of liquor thus employed must, therefore, amount to 121,937,457 ÷ 700•56=174,057 gallons.

(b) Feints from the low-wines still. The condensation and cooling of two fractions (strong and weak) of feints, either from the low-wines or low-wines and feints mixed as previously explained, to the temperature 80°F., require a quantity of water which may be calculated as follows:

The 4,762 gallons of such mixed feints have an average weight of 9•137 lbs. per gallon at 60°F. The weight of such feints is thus 4,762 x 9•137 lbs.=43,510•3 lbs.

The heat-units to be absorbed in the condensation alone of these distillates amount to 4,762 x 9•137 x (965°F. x •7)= 43,510•3 x 675°•5F.=-29,391,207.

To these must be added the heat-units to be absorbed in the cooling down to 80°F.—namely, 4,762 x 9•137 x ·9 x (201-80)=43,510•3 x 121°F. x •9=4,738,271.

Hence the total number of heat-units to be absorbed in the combined condensation and cooling of these feints amounts to 34,129,478.

These distillates are cooled like the low-wines, and with the same supply-liquor. Hence the thermal units absorbed per gallon of such liquor is as before-700•56. The quantity of liquor must necessarily be 34,129,478÷700•56=48,717 gallons.

(c) The final condensation and cooling of the strong and weak fractions of .feints, and of the spirits from this third distillation to the temperature 62°F., require a quantity of water which may he calculated as follows:-

The 4,200 gallons of spirits and final feints have an average weight at 60°F. of 8•948 lbs. The total weight of such distillates is, therefore, 4,200 x 8•948 lbs.=37,5811bs.

The heat-units to be absorbed during the condensation alone of these distillates amount to 4,200 x 8•948 x (965°F. x •63)= 37,581 x 607°•9F.:=22,845,489•9.

To these must be added the heat-units to be absorbed during the cooling down to 62°F. of the liquids in the spirit-worm—namely, 4,200 x 8•948 x •84 x (195— 62) =- 37,581 x •84 x 133°F. =4,198,549

Hence the total number of heat-units to be absorbed in the combined condensation and cooling of these final distillates amounts to 27,044,038.

The distillates are cooled to 62°F. by the supply-liquor at 45°F. To maintain a mean temperature of 62°F. in the tank or tub the liquor must perforce flow over at 79°F., so that each supplied gallon weighing 10•008 lbs. absorbs 79— 45°F.=34°F. x 10•008=-340•27 thermal units. The quantity of liquor used in this tank must necessarily be 27,014,038 ÷ 340•27=79,478 gallons.

Collecting into one total the three quantities of liquor used *in* the respective tanks, we have thus—

(a) 174,057 gallons, *(b)* 48,717 gallons, *(c)* 79,478 gallons; total, 302,252 gallons.

This quantity is equivalent to a trifle over 6 gallons of water for the whole process on each of the 50,000 gallons of wash tendered for distillation.

A weekly supply of this amount, 302,252 gallons, must therefore be provided for refrigerating-tank purposes if this simple but wasteful system of refrigerating the distillates be carried out.

2nd Example — Pot-still distillation. But extravagant as the above system of refrigeration appears to be, it is positively economical when compared with a newer system, in which all the three worms are fitted into one general or composite tank without independent compartments, supplies, and overflows. For, as the spirit, let it be supposed, must be cooled down to 62°F., it becomes necessary, so long as that still is working, to cool down both the other distillates to the same low-collecting heat, although they are to be run direct from the receivers to the stills to be immediately heated up again. The additional time and fuel expended in again raising several thousand gallons through an additional 18°F. can be better ascertained by actual experience than they can be demonstrated by an elaborate calculation. The cooling through the additional 18°F. means that for all three distillates the overflow must be regulated for 79°F., instead of at 115° for the two larger distillates. Without going into details, but applying the data employed in the previous example, it would appear that by such a system, instead of some 302,252 gallons of refrigerating liquor expended weekly, the cooling of the low-wines and of the succeeding distillates to the same temperature as the spirits would require an extra quantity of practically 246,000 gallons, or a total quantity of 548,000, instead of 302,000 gallons, weekly. That is to say, 11 gallons of liquor per 1 gallon of wash distilled, are required instead of only the 6 gallons of the previous example. Whilst an effective division of such a composite worm-tank into three really independent compartments with their supplies and overflows would save all this enormous quantity of liquor, something might, indeed, be set off if the three worms were located or distributed at particular distances, owing to incomplete convection of heat; but the wisdom of completely isolating them is quite apparent.

It is now open to us to consider the more economical methods of condensation and cooling practicable at pot-still distilleries, and to draw instructive comparisons of these with those systems which are usually adopted. The comparison may be confined, at present, to the example already quoted—namely, of a weekly output of 50,000 gallons of wash, with treble distillation, and the final collection of feints and spirits as therein given.

As may have already been inferred, the greatest economy in the refrigerating liquor can be introduced, not by merely utilizing the advantages of a colder liquor-supply, nor even limiting its expenditure by raising the collecting heats of liquids in the receivers by a few degrees, *but rather in regulating the outflow from the principal refrigerating tanks, and then subsequently finishing the cooling in supplemental tanks to the desired point.* It becomes possible in this way, moreover, to utilise a second time the overflow from the spirit-worm tank or compartment, and to do so immediately, without the delay of a night's exposure in the open air in ponds or shallow tanks.

A few representative cases will best exemplify the great economies which it is easy to effect upon the foregoing systems.

3rd Example—Pot-still distillation. In this case the neck of the wash-still, before the worm proper begins, is made to pass, after the first horizontal bend, through a preliminary tank before the principal tank is reached. In this vessel, which is fed independently with the cold liquor-supply, or which may receive the overflow from the spirit-worm tank, the heated vapours are partly cooled and condensed before they pass lower down the worm. It is arranged that the overflow from this tank shall be at 175°F., though it could be carried higher still. This overflow is convenient for boiler-feed, cleansing, or other purposes; but, if not otherwise required, can be run back to cooling ponds for future use. The supply, as in the previous examples, is at 45°F.: hence each gallon of such supply, quitting the preliminary tank at 175°F., must absorb at least (175°— 45°) =130°F., and 10•008 x 130 =1301 heat-units, instead of merely 700•56 units for an overflow arranged at 115°F. in the main wash-still worm-tank; but if the overflow be arranged for 185°F., the heat-units absorbed per gallon of supply liquor, weighing 10•008 lbs. at 45°F., would be at least 10•008 x (185•45)=1401•12, this quantity being exactly double the absorption per gallon measure compared with the overflow at 115°F.: hence each gallon of supply liquor so utilised to this higher extent does twice as much cooling or condensing work as two gallons used in the ordinary way. A weekly expenditure of 10,000 gallons, assuming the distillations to be performed weekly, is therefore equivalent to an ordinary expenditure of 20,000 gallons, showing thus a net saving of 10,000 gallons per week on the wash-vapour refrigeration alone.

This example, put into practical shape, would appear as follows :—Condensation and cooling of 12,500 gallons of low-wines, strong and weak, in the wash-still worms down to 80°F., by means of liquor supplied at 45°F.:—

Total heat-units to be absorbed in this operation, 121,937,457 (see page 43).

The heat-units absorbed by 10,000 gallons of water passing through the preliminary tank are:-10,000 x 10•008 x (185 — 45) = 100,080 x 140=14,011,200, leaving therefore 107,926,257 yet to be absorbed in the main tank. But here each gallon only absorbs 700•56 heat-units, so that the quantity of cold liquor yet required is 107,926,257+700•56 =154,057 gallons. To these add the 10,000 gallons used in the introductory tank; the total quantity used in the refrigeration of the low-wines by this method is, therefore, 164,057 gallons, as against 174,057 gallons found necessary in the previous method of condensing and cooling (page 43) ; a saving of 10,000 gallons in one refrigeration, and being nearly 6 per cent. of the total quantity of liquor formerly used in this one operation.

4th, Example—Pot-still distillation. But if fractional refrigeration be adopted at all, there is no reason why it should be limited to this extent. It is evident, from the above example, that the preliminary tank therein supposed to have been used, must have been small compared with the main tank, for merely

10,000 gallons of liquor passed from its overflows during the same time, or in somewhat less time than about fifteen and a-half times that quantity passed through the overflows of the large tank. Confining the attention once more to the wash-still, and to the low-wines in vapour driven off from it, it is plain that the preliminary tank is capable of being increased to a far larger proportional capacity than 1 to 15k, to which ratio the previous example has restricted it. If, then, instead of passing merely 10,000 gallons of liquor though the small tank, and thereby raising such liquor from 45° to 185°F. in its passage, four times that quantity were so expended, the use of 40,000 gallons in this way would, it is evident, be equivalent to the use of 80,000 gallons in the ordinary way, the saving effected being, therefore, 40,000 gallons net from the low-wines refrigeration alone, and every time the stills were started for the period. The extra bends or coils necessary in the preliminary tank would be far more than compensated for by the large saving of copper in the second tank. Besides, there would be a saving of over 24 per cent. of refrigerating liquor upon every distillation of wash. There must necessarily be a practical limit to the capacity of such a preliminary tank, for under any circumstances, with an overflow at 185°F. the average heat of the liquor therein would be too high for directly collecting a distillate in the receivers, being in all cases at least 115°F., or the average between 45° the supply and 185° the outflow. Even 115°F. would be too high, it may be supposed, and is 35° above the 80°F. mentioned in former examples, nor could even that temperature — 115°F.— be reached unless the quantity of liquor passed through such a preliminary tank amounted to 87,028 gallons. This would be exactly half the quantity required for cooling to 80°F. in the ordinary method, or a saving of 50 per cent. of refrigerating liquor weekly in the collection of low-wines, but involving their collection at 115°

instead of at 80°F. This maximum economy in the expenditure of cooling liquor, followed by a further one in the time and fuel necessary for heating up the warmer low-wines, may not be considered practicable, except in warm climates, where it sometimes would be unavoidable. There the supply liquor for cooling might be at 70° or 80°F., unless refrigerating machinery were employed. Nevertheless, it will now be shown that it is quite practicable to save over 25 per cent. of the supply liquor, to say nothing of the weight of copper in the shortened worms, and yet to cool down the low-wines to 80°F., if they must be collected so low.

And here it is convenient to call attention to Newton's celebrated law of cooling. It is as follows :—" The quantity of heat lost or gained by a body in a second, is proportional to the difference between its temperature and that of the surrounding body." This law Dulong and Petit have shown applies strictly when the difference in temperature between the two bodies does not exceed 63°F. Beyond this difference the loss or gain is even greater than the proportional difference.

Newton's law applies particularly to a saving of time, but the chief object of fractional condensing and cooling is to effect the more important saving of refrigerating liquor. It is therefore encouraging to find that both time and liquor can be saved by the application of the law.

5th Example—Pot-still distillation. Suppose the preliminary worm-tank and the second one to be of such capacities, fed at such rates, with the same cold liquor, that the overflows were regulated to register 185°F. for the former and 115°F. for the latter. To cool the distillate (low-wines) to 80°F. there are, as already stated, some 121,937,000 thermal units to be absorbed some way or other. By using the preliminary tank, each gallon which overflows therefrom, does work equivalent to two gallons overflowing the second or an only tank, for each gallon absorbs 1,401•12 heat-units, instead of only 700•56. The net capacities of these tanks, after deducting for space occupied by the worms, could be greatly reduced, as less water will be shown to be required to do the previous work, whilst in this present example the respective capacities of the preliminary and the second tank may be approximately as 1 to 2, so that one-half a given quantity of liquor may overflow from the first as from the second tank. The overflow, it may be remarked, with the same pressure for supply, would be more rapid from the preliminary tank than from the second one if of similar capacities, because 1,000 gallons raised from 45° to 185° become expanded to 1,032 gallons, but to only 1,010 gallons at 115°F. Now in the example described on page 35, it was shown that 174,057 gallons were necessary for the refrigeration of the low-wines alone. If only half this quantity were used in one tank with an overflow at 185°F., then, as it has just been shown in the previous example, the low-wines would be collected at 115° instead of at 80°F. But if it be arranged that the liquor passing in a given time through a preliminary tank be restricted to one-fourth this original quantity—namely, to 174,057÷4=43,514 gallons—this quantity overflowing at 185°, is equivalent to twice that quantity, or 87,028 passing over at 115°F. It thus absorbs 43,514 x 10•008 x (185 — 45) = 60,968,335 heat-units—just half the total number of units required to be disposed of. There is only the moiety to be absorbed in the second tank—that is to say, another 60,968,335 heat-units. Here, each gallon outflowing at 115°F. absorbs, as before, 700•56 units of heat, and as 87,028 gallons x 700•56= 60,968,335, the necessary work is done. Hence the total quantity of liquor requisite in this arrangement is 43,514 gallons *plus* 87,028 gallons; total 130,542 gallons, instead of 174,057. The saving here is 25 per cent. for each working of the wash-still.

NOTE.—As to the expansion of water due to increase of temperature.

The co-efficient of the expansion of water through the range $39^1/_5$°F. to 212°F., is •04315, therefore 100 gallons at $39^1/_5$°F, become 104•315 at 212°F. The mean expansion per degree of temperature is, therefore, $^{.04315}/_{212-39.2}$=•000249. Thus, approximately 100 gallons at 62°F. expand to 100+(•000249 x 100)= 100•0249 gallons at 63°F., whilst they contract to 100—(•000249 x 100)=99•9751 at 61°F.

Attention may now be directed to other causes of the prolific waste in refrigerating liquor and to the consequent inefficient cooling. They operate very frequently. One common evidence of their existence may be seen in the fact that the overflow liquor at its several exits will be several degrees of temperature lower than that of the distillate which is running through the sampling apparatus of the safes at the worm-ends. The cooling liquids within the worms rush through these too rapidly, in consequence of their excessive width and of the deep incline at which they are set. The heat of these liquids, therefore, is not parted with to the surrounding water, so that a supply of refrigerating liquor—say at 50°F.—will overflow at 60° or 62°F., whilst the distillate is issuing from the worm-ends as high as 65° or 68°F. Unsatisfactory working similar to this may be noticed in shallow or in deep tanks, in principal or in supplementary ones. The cause in most cases may be found under one or more of the following headings :—

(1.) In the too rapid supply of cold liquor below the worm, and at too high a pressure, the overflow being hastened accordingly,

(2) In the excessive size in respect of internal diameter or bore of the worms, so that the liquids within merely come in contact with a small portion of the internal surface, and flow rapidly forward to the discharge.

(3) In the absence of the syphon-shaped regulator from the worm discharge, so that the flow is unchecked.

(4) The too rapid sloping or setting of the worms in the tanks.

(5) The fact that the worms are thickly coated with a deposit of mineral or vegetable matter from the water-supply. This effectually diminishes the conduction of heat from within. So, too, the accumulation of oily matter inside.

(6) The faulty distribution of the coils in the tanks. or of the supply liquor at the bottom of the tanks, the cooling consequently being unequal.

(7) And, generally, from adherence to the employment of circular coils of larger diameter in preference to those of flat or elliptical-shaped coils of smaller diameter, or of a number of parallel tubes of small bore.

Little need be added, by way of explanation, to the above catalogue of the causes of waste of liquor. The extent of such waste is underrated. To promote rapid cooling with a minimum quantity of refrigerating liquor, the greatest possible surface or area of the hot liquid should be brought constantly into contact through the medium of clean copper with the cold liquor without. The coils of the worms should be so distributed, and the cold liquor supply so spread, that, as near as possible, an uniform temperature may prevail at any particular depth or stratum of the cooling liquor. Every gallon of the supply liquor should be utilised to the same extent, so that on reaching the surface the temperature be practically uniform at all the overflows. The advantage of flat or elliptical-shaped worms over circular ones is, that the contained liquid, throughout its run, is brought nearer to the surrounding cold liquor, and without any increase or alteration in the amount of copper required for construction. This may be seen by comparing an ellipse with a circle, taking figures in which the respective areas and perimeters of each sort are identical in magnitude. Not only will less water suffice, but time will be saved, and probably the worm could be abbreviated. The advantages of substituting for worm-coils of great bore an appropriate number of smaller ones must also be recognised here. Nor can this subdivision begin too soon, but may be applied to the first portion of the worm just below the first bend from the neck, and whilst vapours—not liquids—are still passing through. It can then be continued throughout the whole length. In this way the neck may be first branched into two or more parts ; further down each branch may be again subdivided, and these branches can, at any required point to suit the exigencies of the tanks or the transference from one to another, be re-united. The final portion of the worm, instead of consisting of one circular pipe 2 to -3 inches in bore, may consist of four, six, or eight parallel tubes, which in their successive bends feed into, and are fed by, distributing boxes at the tank ends. Instead of one pipe 3 inches in diameter, the same weight of copper could be utilised for making three parallel tubes 1 inch .in diameter; or, instead of one length in the middle part of the worm and of 6 inches diameter, three parallel lengths each of 2 inches diameter. These variations simply mean the use of equal weights of copper, but bring about a far more rapid transfusion of heat from within, and therefore accelerate the cooling. Though the distillates in these proposed cases would be compressed within a smaller area or volume, yet the heat would be parted with more rapidly than through the wider tubes only half-filled. An objection to this subdivision may be made on the ground of the additional trouble of cleaning or flushing out the worms; but it is only a minor objection, and could be overcome by opening the feeding-boxes which hold the parallel tubes, and by the use of flushing cocks on the worms.

One or two more examples of economical condensation and cooling now follow.

6th Example—Pot-still distillation. In the example given on page 42, and which comprised three distinct distillations and condensations—namely, of wash into low-wines, of low-wines into feints, and of feints with low-wines into the final spirits and feints—it was shown that, with the heats of collection into receivers there adopted, a total of 302,252 gallons of liquor supplied at 45°F. was necessary for dealing with 50,000 gallons of wash—that is, a little over six gallons of liquor for each gallon of wash fermented. Considerably more than half the quantity—namely, 174,057 gallons—was used in the condensation and cooling of the low-wines alone. It was subsequently shown on pages 51 and 52, that by a practicable arrangement this quantity of liquor expended upon the vapours from the wash could be reduced to 130,542 gallons, so effecting a saving in a single operation of 43,515 gallons, or of 25 per cent. This reduced the total for the three distillations from 302,252 to 258,737 gallons of liquor. In this way, instead of an expenditure of over six gallons of refrigerating liquor for each gallon of original wash, the reduced quantity used would only be a little over five gallons. But it is evident that the original quantity of refrigerating liquor could be still further reduced if the same economical arrangement were applied to the condensation of all the feints and spirits vapours, instead of restricting it to the low-wines vapours from the wash-still. Even with the vapours from the wash a further reduction in refrigerating liquor could be carried out, which would save also much time and fuel spent in heating up the wash-still itself, by leading the neck of the wash-still, for one or two bends, and before the first refrigerating tank is reached, through the wash-charger first. This plan would not only raise the heat of the wash just passing on to the still, but would enable the preliminary tank already described to be worked at a lower overflow, and without the use of any extra liquor therein. What is more to the purpose, it would allow of the same temperature of overflow—say 185°F.—but with the expenditure of far less water. The saving in time and fuel would be effected with all the wash except the first charge of the still, and so long as any wash remained in the charger to cover the bends of the still-neck. This plan is quite practicable if the wash-charger be in an elévated position.

7th. Example—Pot-still distillation. But even if this plan be not applicable, there remain the constant methods either of using a supplementary or a preliminary refrigerating tank. Now suppose that not only the low-wines vapours, but also all the feints and spirits vapours be conducted through the same or through independent preliminary tanks arranged for overflowing at 185°F., there would then be a saving in each of the three distillations; that from the wash distillation has already been shown to amount to 43,515 gallons of refrigerating liquor.

Then as to the first feints collected from the low-wines still. On page 44 it has been shown that the heat-units, to be absorbed in the condensation and cooling of these vapours, amount to

34,129,478, requiring, by the usual system, some 48,717 gallons of water for the refrigeration of the distillate from 12,500 gallons of low-wines—that is, nearly four gallons of water for each gallon of low-wines dealt with. Now if, instead of using only one tank or tank compartment for performing the work of reducing all the distillate to 80°F. before it is allowed to enter the feints receivers, there be a preliminary tank overflowing at a far higher temperature (let us say 185°F. again in this case), then each gallon of liquor so overflowing would take with it 70°F. more heat. This expressed in pound heat-units—10.008 lbs. being the weight of one gallon of liquor at 45°F.—would mean 10•008 x (185°- 45°) = 10•008 x 140 = 1,401•12, instead of only 700•56 heat-units. As in the case of the improved wash-still worm-tank, therefore, each gallon of supply performs double the original work previously expected from it. Thus if out of the original 48,717 gallons of liquor set apart for the refrigeration, one-fourth were passed through the preliminary tank so as to overflow at 185°F., instead of at 115°F.—that is, if 48,717÷4=12,179 gallons were so used—they would be equivalent to just double that quantity as formerly employed—that is, to 24,359 gallons. Only the remaining 24,359 gallons would now be wanted to complete the work in the main tank. Hence there is a direct saving in this single operation of 12,179 gallons, or of 25 per cent. And thirdly, a similar economy could be practised in respect of the final spirits and feints collected from the feints distillation. In this operation the number of heat-units requiring to be absorbed is 27,044,038 (page 45). But instead of an overflow of refrigerating liquor at 79°F. from one tank, any portion set apart for use in the preliminary one could be caused to overflow at a far higher temperature—and let us adhere to 185°F. still—for this purpose. Then each gallon of supply liquor, instead of absorbing only 79°—45°=34°F., would absorb 185°—45°=140°Fahr., or more than four times as much, each gallon therefore absorbing 1,401•12 heat-units, instead of only 340•27, the exact ratio being 4•1 to 1. The original quantity of liquor used in this operation was calculated to be 79,478 gallons. Suppose one-fifth of this quantity—namely, 15,896—were relegated to a preliminary tank, or to the one composite preliminary tank used by all three worms. As this quantity so used does 4·1 times the former amount of refrigerating work, it actually takes the place of 15,896 x 4•1=65,172 gallons, so leaving only 14,306 gallons necessary to complete the cooling down to 62°F. in the main tank. The saving in this operation is 79,478-30,202=49,276 gallons.

The respective quantities of refrigerating liquor saved in the three distillations are, therefore :—

(1) 43,515 gallons, (2) 12,179 gallons, (3) 49,276 gallons, giving a total saving of 104,970 gallons out of the original outlay, 302,252 gallons, leaving only 197,282 gallons, as required by the improved system of refrigeration. More than one-third of the gross quantity, therefore, is saved each week or each period, reducing the average quantity of refrigerating liquor required per original gallon of wash from six to less than four gallons.

8th Example—Pot-still distillation. There are other equally practicable methods of effecting a large saving in refrigerating liquor. Thus, in cooling spirits and strong feints for final collection to, say, 62°F., the main tank can be utilised to effect the cooling down to 80° or 90°F., and a small supplementary one can finish the work. Thus, in the main tank the overflow can be arranged for 120°F., and in the additional one for 79°F. to effect this object of a collecting temperature of 62°F. in the receivers. In such a case, each gallon of liquor overflowing at 120°F. carries away 120° — 45°=75°F.; or, in heat-units, 10•008 x 75°=750•6, instead of only 340•27 if the whole overflow were set at 79°F. Each gallon, therefore, would do the refrigerating work of 750•6÷340•27=2•2 gallons of the original water. Hence, if of the total 79,478 gallons set apart previously for refrigerating this third distillate (see page 45), one-fourth of it, or 19,369 gallons, were allowed to overflow in the first tank at 120°F., the work performed thereby would be equivalent to that formerly accomplished by 19,369 x 21=42,612 gallons, leaving, therefore, only 79,478 — 42,612=36,866 gallons necessary for finishing the cooling to 62°F. The saving so effected would be 42,612—19,369= 23,243 gallons, being considerably over 25 per cent. of the 79,478 previously used.

An important point must be elucidated here in respect to condensing and cooling. In examining the details given on pages 42 to 45, it will be seen that in each refrigerating operation, the heat-units to be absorbed in the condensation of the vaporous distillates to the liquid form, are from. 5¼ to 6½ times as numerous in those examples as the heat-units to be absorbed in the subsequent cooling down of the liquids. This is due to the vast amount of latent heat concealed in the respective vapours, and not to their temperature, so that 5¼ to 6½ times as many gallons of liquid at any temperature are consumed in absorbing latent heat as in effecting cooling when overflows are constant in temperature. Now, in a well-constructed worm-tank, with the worms properly distributed therein, if liquids only were passing through the worms, the temperature at which the liquids in such worms discharge is practically identical with the mean temperature of the cooling liquor outside the worms, and that is halfway between the entering and overflowing heat point. Not so with vapours, however, for, although the temperature of the condensing liquor is raised very rapidly, it is a long time before any reduction in the temperature of the vapours occurs, the potential energy of the condensing liquor being expended in absorbing the latent heat as this is gradually delivered up. Thus, notwithstanding that the inlets and overflows of the condensing liquor are arranged for 45°F. and 185°F. respectively, and the mean temperature of such liquor be $(45°+185°)/2$=115°F., yet the mixture of vapour and liquid in the worm will be still at a far higher temperature than this mean, however well the worm may be distributed and surrounded by the refrigerating liquor. The cooling-down work is mere child's play compared with the previous condensing work. The maximum economies that could be effected would be by performing all the condensation in the first tank and all the cooling in the second, or second and third, to the limits of practicability. An engineer or distiller can be guided in this connection by a reference to the boiling-points of the usual distillates, an abridged table of which is given on page 42. If he can construct his worm-tanks so that all the vapours be condensed in the first one and the resulting liquids flow on through a continuation of the worms into the cooling tank, or the first cooling tank, as distinct from the condensing tank, he will have effected the greatest possible economy in refrigerating liquor, and ought to have done so, too, in size of worms and tanks. The calculations for finding the necessary quantities of water for each tank are not difficult. By referring to the small table, it is seen that the low-wines collected boil off the wash at 208°F. They therefore condense under an equal pressure at the same temperature, but they must first part with 105,663,881 heat-units. Now with an overflow from the tank arranged for 200°F.—that is, 8°F. on the safe side of the liquefying point of the low-wines vapours—each gallon overflowing would carry off 10•008 x (200°— 45°)=10•008 x 155°=1,551•24 heat-units, so that, to complete the condensation, 105,663,881+ 1,551•24=68,115 gallons would be required. All that remains to be done is to cool the liquid from 208° to 80°F. Now with one tank overflowing at 115°F. each gallon conveys away 700•56 heat-units, and as there still remain 16,273,576 of these in the hot liquid, the remaining absorption is completed by 16,273,576÷700•56=23,229 gallons. The total quantity used in the refrigerating work is therefore 68,115 + 23,229 = 91,344 gallons. These are in place of 174,057 formerly used, so saving 82,713 gallons, or nearly 48 per cent. on the single operation. Even this is not the maximum saving of liquor, as two smaller tanks could be used for consecutive cooling, instead of one larger vessel. In this case the former tank would be arranged to overflow at 150°F., and the final one at 115°F., as usual. The liquor in the former case carries off 105°F. of heat, and in the latter 70°F., as previously. Thus each gallon issuing from the former performs the work of 1½ gallons from the latter. Say 5,000 gallons were devoted to the former tank, they would do the work of 7,500 issuing from the latter, saving thus 2,500 additional gallons. Adding this saving to the larger one in the condensing tank, the gross saving amounts to 85,213 gallons, out of 174,057 used in the older system, and is 49 per cent. It is hardly necessary to point out once more that this enormous saving of liquors in condensing the vapours can only be reached when the worms are suitably branched and contracted so as to prevent the vapours from rushing through them unscathed, that baffles and feeding-boxes may be necessary, and the air-pipes be placed much lower down, and that at suitable bends there may have to be escape pipes for leading back or forward any portion of liquid collecting in the worms, and that the flow must be regulated by syphon-pipes. As exactly the same method of practical refrigeration can be applied to all three distillations, it appears that the quantity of refrigerating liquor necessary for finishing off treble distillates from each gallon of original wash can be reduced from 6 to 3 gallons.

9th Example—Patent-still distillation. As applied to patent-still distillation, the calculations for finding the quantities of cooling liquor are much shorter, as there is no latent heat to be taken into account. In the examples on pages 276 and 277, *Manufacture of Spirit,* it has been shown that to collect 1,000 gallons of strong spirit at a temperature of 62°F., and at 65 over proof, and with a cooling supply at 50°F., the process, without fractional cooling, required 2,580 gallons of liquor, but only 1,433 when fractional cooling was introduced. Through a slight error in that calculation the latter quantity was inserted instead of 1,241 gallons, so that the saving in cooling 1,000 gallons of the strong spirit by fractional methods was really 1,347 gallons of liquor, or of 51 per cent.

The following example, compiled with a few minor corrections of standards, will serve to show the saving that can be effected in the patent-still distillation methods of cooling :—

The specific heat of spirit at 65 over-proof and of •8191 specific gravity at 60°F. is •65 (water=1), and not as in the example just referred to. The spirit quits the spirit-plate, say, at 150°F., and is collected in the receivers at 60°F., being cooled by water flowing into the tank at 50°F., and overflowing at 70°F. The heat-units to be absorbed are, therefore, 8•191 x 1,000 x •65 x (150•60)=479,209•5. The water at 50°F. weighs 10•007 lbs. per gallon; each gallon absorbs 20°F. heat before overflowing: hence the heat-units absorbed per gallon are 10•007 x 20= 200•14, so that the total quantity of liquor required would be 479,209•5÷200•14=2,394 gallons, or about $2^2/_5$ gallons for each bulk gallon of spirits collected at 65 o.p.

10th Example—Patent-still distillation. Now in the more economical case of fractional cooling, say the first tank overflows at 120°F., and the second at 70°F., as before. Each gallon passing from the first tank absorbs 120°F.—50°F.=

70°F., or in heat-units 70 x 10•007=700•49, instead of 200•14 only. The total heat-units tendered for absorption are the same—namely, 479,209•5. The mean temperature of the first tank must be $^{(320°+50°)}/_2$=85°F., if inlets and overflows are worked properly. This mean temperature—85°F.—will be the temperature of the distillate as it runs from the worm of the first tank, if the flowing out be properly checked by a syphon-pipe. Hence the heat-units to be absorbed in the first tank alone are 8•191 x 1,000 x •65 x (150-85)=346,095•75, which number divided by 700•49 gives 494 gallons of liquor required for the work. There remain then only 479,209•5—346,095•75=133,114 heat-units, to be absorbed in the second tank. As the quantity absorbed per gallon of liquor is there reduced to 200·14 heat-units, the remaining quantity of liquor to be used will amount to 133,114±200•14=665 gallons. These added to the 494 already used in .the first tank show a total employed of 1,159 gallons, to be set against the 2,394 gallons used when fractional cooling was not adopted. The saving is, therefore, 1,235 gallons, equivalent to 51 per cent. of the gross quantity formerly expended. This quantity is saved upon every 1,000 gallons of strong spirit collected, and is, indced, an enormous economy when resort has to be made to a lower supply.

Quantity of Refrigerating Liquor used in the Cooling of Wort.

The quantity of cold liquor required for finishing off the cooling of the wort running from the coolers, and other wort-collecting vessels, on to the refrigerators, and thence to the fermenting wash-backs, can be readily calculated with approximate accuracy, if a few details of the operation be known. The quantity is small compared with that required in the subsequent condensing and cooling processes after distillation, and is small also compared with the quantity required for similar cooling operations at a brewery. For at breweries, the original collecting temperatures of the wort discharged from the coppers are far higher, whilst the cooling has to be carried down some 10 or 15 degrees of temperature lower than those which are generally considered necessary at distilleries.

In distillery brewings, the cooling of the wort from the mash-tun draining temperatures to the initial fermentation temperatures, can be carried out quickly. These draining temperatures range from about 140° to 150°F. for the wort passing on to the coolers, varying within, or slightly above or below these limits, according as there may be one mash only, or more consecutive mashings of the same materials.

To take a representative case, in which 50,000 gallons of wort are collected in each period, and are run on to the coolers at due intervals, and at temperatures ranging from 140° to 150°F. After collecting there, the wort is very soon fed on to refrigerators, though the cooling can be, and frequently is, finished without the aid of refrigerators, being performed by mere exposure on shallow coolers, aided by revolving fans. But assuming the use of refrigerators, and that the average temperature of the wort when entering these vessels is as high as 130°F. The final temperature when the wort quits these vessels may be taken as 70°F. There is thus a loss of 60°F., chiefly caused, but by no means entirely so, by the action of the cold water, which enters the lower end of the perpendicular vessels or the remote end of the horizontal ones.

Assuming, for the present, that this loss of temperature of 40°F. by the wort is due to the heat abstracted from it by the cold refrigerating liquor; further, that the specific heat of wort is the same as that of water, though it is probably less, then the heat-units parted with by the wort and assumed to be entirely acquired by the water, would for 50,000 gallons cooled from 130° to 70°F. be calculated as follows :—

Weight of a gallon of water at 70°F., 9•991 lbs., and though the wort may have a specific gravity of 1,045 to 1,050 at 60°F., and consequently a weight per gallon of 10•439 to 10•488 lbs. at 70°F., yet these increased weights must be ignored. For the excess weight per gallon over the 9•991lbs. is made up of malt-extract or of similar substances, and these have a lower specific heat than water, besides which, they displace a portion of the water in each gallon. Therefore, the wort may, so far as its specific gravity or as its weight per gallon is concerned, be taken as water, whilst its specific heat per given measure or per given weight, though taken as equal to that of water, would really be less, thus requiring less cooling water for reducing its temperature.

Take therefore at 70°F. the weight per gallon of wort as 9•991lbs. irrespective of its gravity.

The heat-units to be parted with by the whole bulk are therefore 9•991 x 50,000 x (130°-70°F.)=29,973,000.

Next, as to the cold liquor supply and its subsequent discharge at the further end after having performed its work. As has been shown so conclusively in the case of condensing and cooling of distillates, the temperature at which this discharge is allowed to occur is the chief governing factor of the quantity of cold liquor necessary. According to the pressure in the cold main, the difference between the temperatures of the entering and discharging liquor may be 10, 20, 30, or 40 degrees Fahrenheit, and possibly a higher difference than 40°F. may be attained, that is to say, when dealing with wort supplied at 130°F., running off at 70°F., cooled by liquor supplied at 50°F. If the pressure be so great that the cold supply only gathers 20°F. in temperature during its course, then double the quantity of liquor will be required, compared with that necessary when a discharge has gathered 40°F. during its course.

In the case being dealt with, let the liquor supply be at 50°F., and the discharge at 80°F Each gallon of cold supply at 50°F., and weighing 10•007 lbs., absorbs 10•007 x (80°—50'F.)= 10•007 x 30 = 300•21 heat-units. The total heat-units tendered for absorption are 29,973,000, hence the quantity of cold liquor required would be 29,973,000÷300•21=99,840 gallons, that is, an average of 2000 gallons per 1,000 of wort.

If the discharge took place at 90°F., the quantity of cold liquor required would be proportionately reduced inversely to the increase in the number of degrees of temperature gained on the whole number taken up. Thus the gain in temperature would be 40°F. instead of 30°F., and each gallon of cold liquor would therefore absorb 10•007 x 40= 400•28 heat-units, and 29,973,000÷400•28=74,880 gallons. This would only involve an average expenditure of 1500 gallons of cold liquor per 1,000 gallons of wort.

It must be pointed out that these gross estimates would be in excess of the quantities of cold liquor actually necessary. Hot wort, exposed on a refrigerator, and especially in the fine films which course dawn the perpendicular ones, would lose an enormous amount of heat in a short time by the natural processes of evaporation and radiation. If the vessel were in a cold, exposed, and draughty position, the surrender of heat would be especially large, particularly if the surrounding air were cold and dry, and were constantly renewed. These conditions would greatly reduce the calculated or gross estimate.

Nevertheless, these estimates are valuable as a guide to both distillers and brewers.

Other Methods of Distilling, Rectifying, Condensing, and Cooling.

All methods of distilling which deliver only liquids into the worms may, for the present purpose, be regarded as modifications of Coffey-still distillation. Throughout the foregoing pages it has been assumed that the operations are conducted either by ordinary pot-still methods or by a system similar to that of Coffey.

Looking ahead for probable improvements in stills and in distillation processes, judging too from the systems which prevail all over the Continent, in the States, in the Indies, and in the Colonies, it is impossible to doubt that methods of rectification accompanying continuous distillation must sooner or later be attempted in connection with our own pot-stills, whether heated by fire or steam. And if not for all the pot-stills, then for the wash-still or for one of the others.

It is only necessary here to refer to such arrangements so far as they effect cooling and condensing and the appropriate quantities of cold liquor therefor.

Thus, suppose the vapours from a wash-still to pass through one or more rectifying vessels before they enter the worm or other condensing plant substituted for it. As a rule, these rectifying vessels effect partial or complete condensation of the low-wines vapours, and do so chiefly by means of new wash, which is then on its way towards the wash - still. The refrigerating liquor has merely to complete the condensation and to effect the further cooling down to the collecting temperature—usually it has only the latter function to fulfil. This arrangement at once effects a great saving in the expenditure of liquor and in the size of the worm. But in this arrangement of a rectifying pot-still, if distillation of the wash is to be continuous, and conducted without the loss of time spent in discharging, recharging, and partially heating-up anew, only one cold charge, the first charge, could be run into the wash-still from the chargers or intermediate chargers. The next and subsequent charges of wash, when the first boiling has exhausted the first charge of all its alcohol, and has driven it over to the rectifying vessels, are therefore supplied hot, having passed through the tubes, coils, or jackets in one of the various forms of feed-heaters or rectifying vessels. The spent-wash in the body of the still will hereafter discharge itself automatically at the proper level. A supply of wash must always be at hand in the intermediate charger, ready to be passed through the tubes, coils, &c., of the rectifying vessels, until all the wash of the period be exhausted, and then cold water will take its place till the close of the distillation. There would be the usual arrangements as in ordinary pot-stills and Coffey-stills for returning to one of the chambers of the column above the pot-still any portion of first-shots which may condense and collect in the rectifying vessels.

But it is seldom that only low-wines or feints are collected from a still of this description, working continuously and fed and discharged constantly until the whole bulk of wash is disposed of. This of course could be done if the object were to work off large quantities of wash, distil them into low-wines, and feed other stills with these spirits, so saving time in these latter stills. Usually the distillation would be completed in the one still by similar working to that of the Coffey-still, that is, without any intermediate condensation, cooling, or reheating of the low-wines. Such a still, with its accompanying rectifying vessels, would begin and end the operation, distilling over consecutively low-wines, feints, and spirits, delivering over the latter in the liquid form to the worm, and finishing by delivering over feints in the same condition.

Pot-still distillation, however, notwithstanding the fact that it is the most extravagant system of distillation possible, has been so long in favour at the whisky distilleries of the United Kingdom, is so popular amongst consumers, and is so enveloped in tradition and romance, that the engineer who attempts to improve upon it at home receives but small encouragement. If, indeed, modifications of pot-stills are introduced, there is a strong disposition to retain the old form of pot-still body, and to attach either a column of rectifying chambers above it instead of the usual neck, or else to attach the rectifying vessels to the neck. These rather than resort to close steam vessels of different shape and construction. For illustrations of the former descriptions of stills, see pages 129 to 131, *Manufacture of Spirit*, and pages 146 to 149 of the *same* work for illustrations of the steam stills.

How long this adhesion to the old methods may last cannot be foretold. Proprietary distilleries will probably remain faithful longest to the old plant, but as the large distilling firms and companies become assured of the extravagance of the system, efforts will be made, in spite of the well-deserved character of old pot-still distillates, to introduce more economical systems.

It has been said already that pot-still distillation is the most extravagant system extant. The foregoing examples showing the necessary quantities of refrigerating liquor, form one strong proof of the allegation, and many other proofs suggest themselves in support of the change. It is not merely that two, three, or four condensations and coolings are necessary, and in which enormous quantities of latent heat have to be

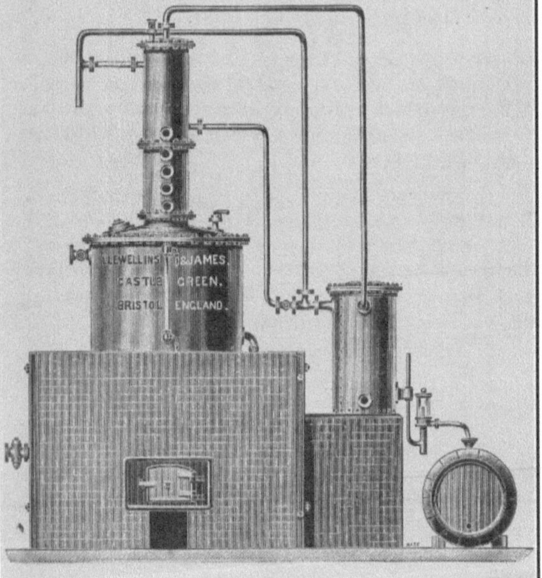

PLATE I.—FIRE-STILL, WITH RECTIFYING COLUMN AND COMBINED FEED-WASH HEATER AND CONDENSER.

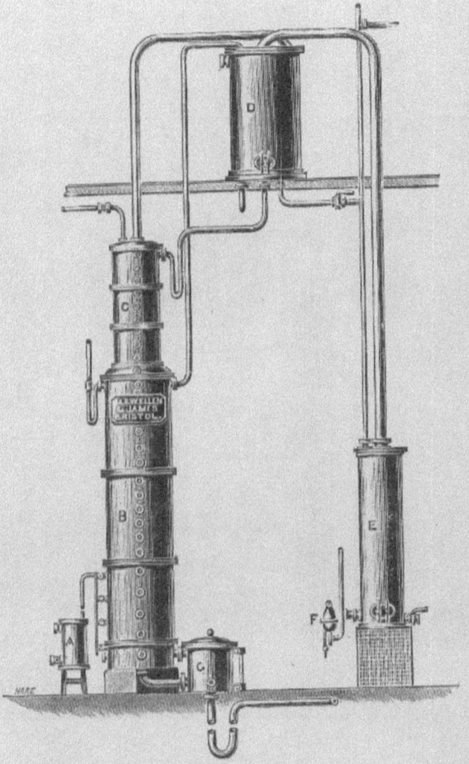

PLATE II.—STEAM-STILL, WITH RECTIFYING COLUMN, FEED-WASH HEATER, AND ADDITIONAL CONDENSER.

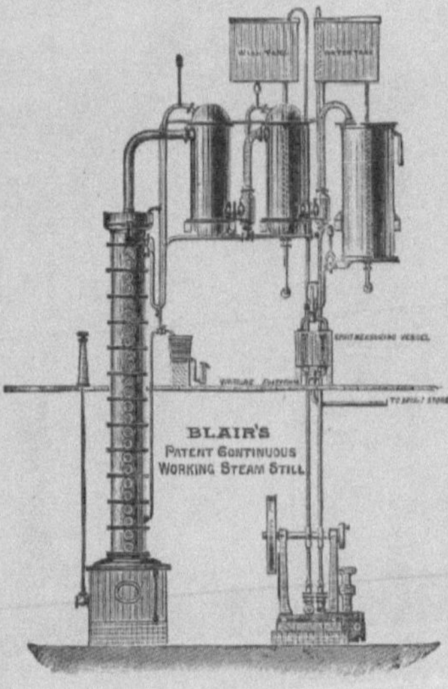

PLATE III.—STEAM-STILL, WITH COMBINED DOUBLE-FEED WASH HEATERS AND RECTIFIERS, AND ADDITIONAL CONDENSER.

rendered sensible, and then disposed of, but two, three, or four heatings-up with fresh fuel are also necessary. Then there must be two, usually three, stills with furnaces or steam-supplies, appropriate worms and tanks (the former made of expensive copper), as well as safes, receivers, chargers, and pumps to collect and deliver the series of distillates. There is, further, the time lost in charging, discharging, and re-charging all the stills, and, finally, there is the fearful waste of potential energy in the form of sensible and latent heat possessed by the vapours, which energy could be so easily directed towards heating up the raw liquids passing forward for distillation.

As to Condensing and Cooling Plant, enough has been inferred from the previous pages to convince one that the ordinary worm arrangements, in spite of some modern improvements, are about the crudest that could be imagined. The modern improvements consist of some slight alterations in the curve or shape of the coil from the circular to the elliptical; in the addition of return pipes from the neck-bend to the stills; in the openings for cleaning purposes; in the arrangement, distribution, and subdivision of the coils ; in the occasional substitution at the right place of parallel tubes for the coil; in the superior distribution of the cold supply liquor and regulation of the warm overflows; and in the regulation of the internal flow of the distillates by means of syphons and reduced gradients.

There remains a wide field open for inventors in this direction, so that, with an equal or less outlay for copper, superior distribution of heat may be arranged, and escape of heat from within to without may be facilitated and regulated. Better arrangements are urgently required for thorough internal and external cleaning of worms or other condensers.

The following tables will assist distillers in estimating the weight of copper necessary for stills, worms, &c.:—

TABLE IX – WEIGHTS OF COPPER SHEETS.

One square foot, No. 21 gauge	.032 inch thick = 1/32 inch, scant	weighs 1·45 lbs.
" " 20 "	.035 " " = 1/32 " full	" 1·59 "
" " 17 "	.058 " " = 1/16 " scant	" 2·627 "
" " 16 "	.065 " " = 1/16 " full	" 2·945 "
" " 11 "	.120 " " = ½ " scant	" 5·436 "

If a pot-still be assumed to be-a sphere in shape, its area or surface in square feet can be found by taking two dimensions in feet, namely, the mean diameter and the mean circumference. The product of these is the surface area in square feet. Thus—

The mean diameter of three such measurements, two taken horizontally at right angles to each other, and the third taken perpendicularly to the other two, was 8½ feet. The mean circumference of two measurements, one horizontal and the other perpendicular, was 25 feet. The surface area was, therefore, 25 x 8½ = 212½ square feet.

If the vessel were exactly the shape of a hollow sphere, one dimension only, either diameter or circumference, would suffice, as the constant ratio of 1 to 3·14159 would then hold.

The weight in the above example can then be approximately calculated from the area if the mean thickness of the copper be known.

TABLE X – COPPER TUBING, WEIGHTS PER FOOT IN LENGTH.

	Circumference in inches of Tube or Pipe						
	1 inch	1½ inch	2 inch	2½ inch	3 inch	3½ inch	4 inch
	lb.	lb.	lb.	lb.	lb.	lb.	lb.
1/32 inch thickness, full, No. 20	·41	·62	·84	1·05	1·26	1·48	1·69
1/16 " " scant, " 17	·67	1·02	1·37	1·73	2·08	2·43	2·79
1/16 " " full, " 16	·74	1·13	1·53	1·92	2·32	2·72	3·11
½ " " scant, " 11	1·29	2·02	2·75	3·48	4·22	4·95	5·68

TABLE XI – ALCOHOL AND WATER MIXTURES, WITH SPECIFIC GRAVITIES, PROOF STRENGTHS (BRITISH AND AMERICAN), AND CONTINENTAL STRENGTHS.

Alcohol and Water Mixture (specific gravities at 60°F.) Water at 60°F.=1.	Weights percent of absolute alcohol (Drinkwater's alcohol sp. gr. •7938 at 60°F.).	Proof Strengths (British) Sikes' Hydrometer.	American (United States) Proof Strengths. Tralles' Hydrometer.	Continental Strengths (Centesimal alcoholometer) Gay-Lussac's Hydrometer.
		(volume)	(volume)	(volume)
1•0000 (water)	None.	100 underproof	None.	None
•9959	2•20	95 "	5•7 per cent.	2•8 per ct.
•9919	4•65	90 "	11•4 "	5•7 "
•9883	6•85	85 "	17•1	8•6 "
•9850	9•20	80 "	22•8	11•5 "
•9818	11•55	75 "	28•5	14•4 "
•9789	13•90	70 "	34•2	17•3 "
•9760	16•25	65 "	39•9	20•1 "
•9732	18•55	60 "	45•6	23•0 "
•9701	21•00	55 "	51•8	25•9 "
•9671	23•40	50 "	57•0	28•8 "
•9638	25•85	45 "	62•7	31•7 "
•9601	28•35	40 "	68•4	34•6 "
•9563	30•80	35 "	74•1	37•4 "
•9520	33•30	30 "	79•8	40•3 "
•9474	35•80	25 "	85•5	43•2 "
•9424	38•45	20 "	91•3	46•1 "
•9371	41•10	15 "	97•0	49•0 "
•9343	42•50	12•4 "	100%(American Pr'f)	50•0 "
•9318	43•70	10 "	102•7 per cent.	51•9 "
•9257	46•55	5 "	108•4	54•7 "
•9198	49•25	Proof	114•2	57•66 "
•9137	52•05	5 overproof	119•9	60•5 "
•9070	54•95	10 "	125•5	63•4 "
•9005	57•85	15 "	131•2	66•2 "
•8938	60•80	20 "	136•9	69•1 "
•8866	63•90	25 "	142•6	72•0 "
•8793	66•95	30 "	148•3	74•9 "
•8718	70•15	35 "	154•0	77•8 "
•8641	73•40	40 "	159•7	80•6 "
•8560	76•70	45 "	165•4	83•5 "
•8476	80•20	50 "	171•1	86•5 "
•8388	83•70	55 "	176•9	89•3 "
•8293	87•40	60 "	182•7	92•2 "
•8192	91•25	65 "	194•9	95•1 "
•8080	95•30	70 "	194•9	98•0 "
•7946	99•75	74•9 "	200•0	100•0 "
•7944	99•80	75 "	--	--
Drink-water's alcohol •7938	*100•00	75•25 "	--	--
Squibbs' alcohol •7955	--	75•37 "	--	--

*Equal to 99•9028 Squibbs' alcohol by weight, and contains •0972 per cent. by weight of water. See pages 321,322, *Manufacture of Spirit*.

The American (United States) gallon measures less than the British Imperial gallon, for 1 gallon (U.S.) = only •833 gallon (British).

1 gallon British measures 1•2, that is, 1½ gallon (U.S.), therefore, 5 gallons British measure 6 gallons (U.S.).

As to Alcoholic Strengths. The table on page 75 shows American proof spirit to be weaker than British proof, it being in fact equal only to 124 under-proof, British. British proof is equal in strength to the American strength 14•2 over-proof, or 114•2 per cent. Hence 100 gallons measure of British proof spirit are equal in alcoholic value to 100 x $^{6}/_{5}$ x $^{114\cdot2}/_{100}$=137•04 gallons of American proof spirit.

The table also shows that Tralles' alcohol (the American standard) and Gay-Lussac's alcohol (the French standard) are both weaker than the British standard alcohol, namely, Drinkwater's. Also, that Tralles' alcohol and Gay-Lussac's are almost identical in strength, thus—

Gay-Lussac's Alcohol,	sp. gr.	•7947	at 59°F.
Tralles' Alcohol,	"	•7946	" 60°F.
Drinkwater's Alcohol,	"	•7938	" 60°F.
Squibbs' Alcohol,	"	•7935	" 60°F.

Hence—

Gay-Lussac's	Alcohol	contains	•25	per cent.	by weight	of water.
Tralles'	"	"	•25	"	"	"
Drinkwater's	"	"	•0972	"	"	"

Columns 1 and 2 in Table XI. are taken from Stevenson's tables as given in his work, Treatise on Alcohol, and with his kind permission.

TABLE XII.

Corrections to be made in the author's previous work,

The Manufacture of Spirit.

Page 46, line 3, in paragraph 3, *read* "•15 lb." *instead of* "15 lb "

Page 199, line 3, *read* "1001" *instead of* "100•"

Page 213, line 5, read "92°F." *instead of* "99°F."

Page 261, line 3 from bottom, *read* "Acetic Ether" *instead of* "Acute Ether."

Page 265, line 1, *delete words* "of water and alcohol."

Page 277, line 8, *read* "401,359 heat-units" and "668 gallons," *instead of* "*516,033* heat-units" and "860 gallons." For the heat-units to be absorbed in the first tank are 8•191 x •7 x 1000 x (150°-80°)= 401,359,

Page 277, lines 19 and 20, read "668 gallons" *instead of* "860," *and the total* " 1,241 gallons" *instead of* "1,433."

Page 283, last line, the number " 10 " in the divisor should be "20," hence total gallons in the quotient should be " 17,518 " *instead of* "35,037."

Page 297, lines 21 and 23, *read* 185 *instead of* 150, and 185 ÷ 500 = •37.

Page 396, in table at foot of page, *read* "113•49" *instead of* "133•49."

Page 411, 8th name, *read* "Watney & Co., York Road."

Page 412, 7th name, *delete word* "Tullymet."

Marcus Ward & Co., Ld., Printers, Royal Ulster Works, Belfast.

BLAIR, CAMPBELL & M'LEAN,

Scotland Street Copper Works,

GLASGOW, SCOTLAND.

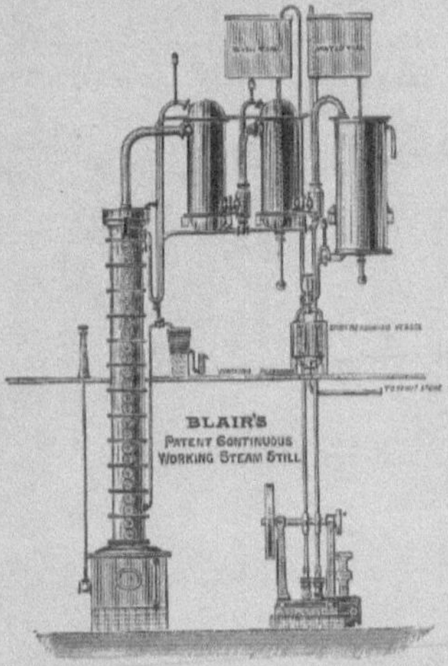

DISTILLING & SUGAR-BOILING PLANT of every description. **"BLAIR'S" PATENT CONTINUOUS-WORKING STEAM STILLS** to produce Rum and Spirits of best quality and strength in one operation, guaranteed to extract all the spirit from the wash. These Stills are the most economical made, both as regards fuel and water.

"COFFEY'S" PATENT STILLS; POT STILLS for direct firing and Steam Jacketed; **VACUUM PANS** in Iron or Copper; Open **SUGAR PANS** or **TEACHES; DEFECATORS; CLARIFIERS; WETZEL PANS;** &c., &c

Speciality—**COPPER & BRASS WORK** of every description.

ESTABLISHED OVER HALF-A-CENTURY. A B C CODE USED.

Telegraphic Address: "BLAZON, GLASGOW." ENQUIRIES INVITED.

Read your whisky.

Made in United States
Troutdale, OR
11/03/2026

56852538R00056